Seminar on
FISSION

Seminar on FISSION

Het Pand, Gent, Belgium,   17 – 20 May 2010

Editors

Cyriel Wagemans
University of Gent, Belgium

Jan Wagemans
SCK•CEN, Belgium

Pierre D'hondt
SCK•CEN, Belgium

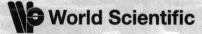
World Scientific

NEW JERSEY · LONDON · SINGAPORE · BEIJING · SHANGHAI · HONG KONG · TAIPEI · CHENNAI

Published by

World Scientific Publishing Co. Pte. Ltd.
5 Toh Tuck Link, Singapore 596224
USA office: 27 Warren Street, Suite 401-402, Hackensack, NJ 07601
UK office: 57 Shelton Street, Covent Garden, London WC2H 9HE

British Library Cataloguing-in-Publication Data
A catalogue record for this book is available from the British Library.

SEMINAR ON FISSION

Copyright © 2011 by World Scientific Publishing Co. Pte. Ltd.

All rights reserved. This book, or parts thereof, may not be reproduced in any form or by any means, electronic or mechanical, including photocopying, recording or any information storage and retrieval system now known or to be invented, without written permission from the Publisher.

For photocopying of material in this volume, please pay a copying fee through the Copyright Clearance Center, Inc., 222 Rosewood Drive, Danvers, MA 01923, USA. In this case permission to photocopy is not required from the publisher.

ISBN-13 978-981-4322-73-7
ISBN-10 981-4322-73-3

Printed in Singapore by World Scientific Printers.

These Proceedings are dedicated to Prof. Dr. Cyriel Wagemans at the occasion of his **65th anniversary**. He was born in Turnhout (Belgium) on 5 January 1945.

Already at the beginning of his scientific career he was interested in nuclear fission, since the topic of his Masters thesis (1966) was a search for the fission isomer ^{242m}Am. Also his Ph.D. and Habilitation theses dealt with nuclear fission. He continued fission studies during his whole career, the main topics being the characteristics of ternary fission and of fission fragment energy and mass distributions in spontaneous and neutron induced fission. He is also an expert in neutron induced fission cross section measurements. Recently he was involved in studies of β-delayed fission.

In 1986 he initiated the "Seminar on Fission" conference series. A scientific highlight in his career was the publication of the reference book "The Nuclear Fission Process" in 1989.

We wish him good luck with his retirement.

Dr Jan Wagemans and Dr Pierre D'hondt
Editors

Fonds Wetenschappelijk Onderzoek
Research Foundation – Flanders

ORGANISING COMMITTEE

P. D'hondt
SCK•CEN Mol
Belgium

M. Huyse
University of Leuven
Belgium

C. Wagemans (chair)
University of Gent
Belgium

SCIENTIFIC ADVISORY COMMITTEE

N. Cârjan
University of Bordeaux, France
NIPNE, Romania

J. Cugnon
Université de Liège
Belgium

F. Gönnenwein
University of Tübingen
Germany

H. Goutte
GANIL Caen
France

F.-J. Hambsch
IRMM Geel
Belgium

M. Itkis
JINR Dubna
Rusia

O. Serot
CEA Cadarache
France

PREFACE

This Seminar is the seventh of a series started in 1986. The first five meetings took place in the historical castle of Pont d'Oye (Habay-la-Neuve), the sixth in the ancient priory of Corsendonk. This time we met in the buildings of "Het Pand", which is a restored Dominican monastery of the 13th century located in the heart of the ancient town of Gent. Seven is a holy number in the Bible, but this seventh Seminar marks the end of my academic career and - unfortunately - at the same time the end of fission research at the University of Gent...

During this meeting, recent achievements in experimental and theoretical fission physics were discussed, giving special attention to low-energy fission and its traditional topics such as fission fragment characteristics, ternary fission, fission neutrons and fission cross sections. Also more specialised topics such as e.g. β-delayed fission, γ-ray spectroscopy of fission fragments and fission properties of super-heavy nuclei were discussed. Furthermore, review papers on 35 years of fission research at the Institute Laue-Langevin (Grenoble) and on mechanisms for particle emission during nuclear fission were presented. Finally, due attention was given to new facilities and detectors.

This Seminar is strongly supported by three organisations: the Fund for Scientific Research Flanders (FWO), the University of Gent (UG) and the Belgian Nuclear Research Centre in Mol (SCK•CEN). The organising committee is very grateful to these sponsors. Also the valuable assistance of the International Advisory Committee, of the Chairmen of the Sessions and of the various people who helped before, during and after the conference is gratefully acknowledged.

I will conclude with a short in memoriam for Prof. Dr. Denis De Frenne who unexpectedly passed away on 6 October 2009. Denis was born on 12 July 1941. He was a member of the International Advisory Committee and took part in the local organisation of this Seminar. He was an expert in photofission and nuclear spectroscopy, but above all an excellent and friendly colleague.

Cyriel Wagemans
Conference chair

CONTENTS

Dedication v

Sponsors vii

Organising and Scientific Advisory Committees ix

Preface xi

Topical Reviews 1

35 years of fission research at the ILL 3
   F. Gönnenwein

Mechanisms for light charged-particle emission during nuclear fission 31
   N. Cârjan

Characteristics of the Fission Process 43

New developments in the microscopic description of the fission process 45
   H. Goutte and R. Bernard

Quantum aspects of low-energy nuclear fission 53
   W. Furman

Phenomenological model improvement for the Monte Carlo simulation 65
of the fission fragment evaporation
   O. Litaize and O. Serot

Excitation energy sorting in superfluid fission dynamics 73
   B. Jurado and K.-H. Schmidt

Systematic analysis of structural effects in fission-product yields and 81
neutron data and the consequences for our understanding of the fission
process and the predictive power of model predictions
    K.-H. Schmidt and B. Jurado

The statistical model in nuclear fission-excitation energy and spin 89
population in fragments
    H. Faust

Investigation of ^{234}U(n,f) as a function of incident neutron energy 99
    A. Al-Adili, F.-J. Hambsch, S. Oberstedt and S. Pomp

Measurement of fragment mass yields in neutron-induced fission of 107
^{232}Th and ^{238}U at 33, 45 and 60 MeV
    V. D. Simutkin, S. Pomp, J. Blomgren, M. Österlund, P. Andersson,
    R. Bevilacqua, I. V. Ryzhov, G. A. Tutin, M. S. Onegin,
    L. A. Vaishnene, J. P. Meulders and R. Prieels

Progress in the atomic number identification of fission fragments 115
    I. Tsekhanovich, A. G. Smith, J. A. Dare and A. J. Pollitt

Fragment characteristics from photofission of ^{234}U and ^{238}U induced 123
by 6.5 – 9.0 MeV bremsstrahlung
    A. Göök, R. Barday, M. Chernykh, C. Eckardt, J. Enders,
    P. Von Neumann-Cosel, Y. Poltoratska, M. Wagner, A. Richter,
    S. Oberstedt, F.-J. Hambsch and A. Oberstedt

Structure effects in the asymmetric fission of ^{118}Ba, ^{122}Ba 131
compound nuclei
    G. Ademard, J. P. Wieleczko, E. Bonnet, A. Chbihi, J. D. Frankland,
    et al., for the E475S collaboration

Constraining fission parameters for highly excited compound nuclei 135
    D. Mancusi, J. Cugnon and R. Charity

Ternary Fission 143

Characteristics of light charged particle emission in the ternary fission 145
of ^{250}Cf and ^{252}Cf at different excitation energies
    S. Vermote, C. Wagemans, O. Serot, J. Heyse, T. Soldner,
    P. Geltenbort, I. Almahamid, G. Tian and L. Rao

Ternary fission studies by correlation measurements with ternary particles 155
   M. Mutterer

Neutron Emission in Fission 163

Prompt neutron emission in ^{252}Cf spontaneous fission 165
   F.-J. Hambsch, S. Oberstedt and Sh. Zeynalov

Energy measurement of prompt fission neutrons in ^{239}Pu(n,f) for 173
incident neutron energies from 1 to 200 MeV
   A. Chatillon, T. Granier, J. Taieb, G. Belier, B. Laurent, S. Noda,
   R. C. Haight, M. Devlin, R.O. Nelson and J. M. O'Donnell

Anisotropic neutron evaporation from spinning fission fragments 181
   L. Stuttgé, O. Dorvaux, F. Gönnenwein, M. Mutterer, Yu. Kopatch,
   E. Chernysheva, F. Hanappe and F.-J. Hambsch

New Facilities and Detection Systems 189

Status of MYRRHA and ISOL@MYRRHA in March 2010 191
   H. Aït Abderrahim, J. Heyse and J. Wagemans

Next generation fission experiments at GSI: short and 199
long term perspectives
   A. Bail, J. Taieb, A. Chatillon, B. Laurent, G. Belier, L. Tassan-Got,
   L. Audouin, B. Jurado, K.-H. Schmidt, J. Benlliure, F. Reymund,
   D. Dore and S. Panebianco

The fission fragment time-of-flight spectrometer VERDI 207
   S. Oberstedt, R. Borcea, F.-J. Hambsch, Sh. Zeynalov, A. Oberstedt,
   A. Göök, T. Belgya, Z. Kis, L. Szentmiklosi, K. Takács and
   T. Martinez-Perez

Digital pulse processing for STEFF 215
   A. J. Pollitt, J. A. Dare, A. G. Smith and I. Tsekhanovich

Fission γ-ray measurements with lanthanum halide scintillation detectors 223
   A. Oberstedt, R. Billnert, J. Karlsson, A. Göök, S. Oberstedt,
   F.-J. Hambsch, X. Ledoux, J.-G. Marmouget, T. Belgya, Z. Kis,
   L. Szentmiklosi, K. Takács, T. Martinez-Perez and D. Cano-ott

Characterization and development of an active scintillating target for 231
fission studies
   G. Belier and J. Aupiais

Fission Probabilities: Barriers and Cross Sections 237

Thermal fission cross section measurements of ^{243}Cm and ^{245}Cm 239
   L. Popescu, J. Heyse, J. Wagemans and C. Wagemans

^{245}Cm fission cross section measurement in the thermal energy 247
energy region
   O. Serot, C. Wagemans, S. Vermote and J. Van Gils

r-process reaction rates for the actinides and beyond 255
   I. V. Panov, I. Yu. Korneev, T. Rauscher and F.-K. Thielemann

High-energy neutron-induced fission cross section of subactinides 263
   D. Tarrio, I. Durán, C. Paradela, L. Tassan-Got, L. Audouin, and N_TOF collaboration

Proton-induced fission cross section of ^{181}Ta 267
   Y. Ayyad, J. Benlliure, E. Casarejos, H. Álvarez, A. Bacquias, A. Boudard, T. Enqvist, V. Föhr, A. Kelic, K. Kezzar, S. Leray, C. Paradela, D. Pérez-Loureiro, R. Pleskac and D. Tarrio

Fission in the Super-heavy Mass Region 271

Formation and decay of superheavy compound nuclei obtained in the 273
reactions with heavy ions
   M. G. Itkis, A. A. Bogachev, I. M. Itkis, G. N.Kknyazheva and E. M. Kozulin

Ternary fission and quasi-fission of superheavy nuclei and giant 289
nuclear systems
   V. I. Zagrebaev, A. V. Karpov and W. Greiner

What can we learn from the fission of the super-heavy elements? 297
   D. Boilley, Y. Lallouet, B. Yilmaz and Y. Abe

Conference Photo 305

List of Participants 307

Author Index 313

Topical Reviews

35 YEARS OF FISSION RESEARCH AT THE ILL

F. GÖNNENWEIN

University of Tübingen, Tübingen, D 72076, Germany

Fission research at the High Flux Reactor of the ILL in Grenoble/France started in 1975 with the installation of the Lohengrin spectrometer for recoiling fission fragments. The separator uses a combination of magnetic and electric sector fields to measure fragment masses and energies with unrivalled resolution. A complementary detector based facility called Cosi Fan Tutte was developed. Highlights were the study of super-asymmetric and cold fission, even-odd effects of charge distributions, systematic investigations of both ternary and quaternary fission, binary and ternary fission induced by polarized neutrons. Exploiting the high neutron flux and the quality of beams having been filtered by curved neutron guides allowed delicate fission cross section measurements to be made serving as references.

1. Introduction

The discovery in 1939 by O. Hahn and F. Strassmann of Ba when Uranium was irradiated by neutrons led L. Meitner and O. R. Frisch to identify and call the new phenomenon "fission". Remarkably, only a few months after the discovery N. Bohr and J. A. Wheeler presented the ground-breaking theory of fission based on the liquid drop model. The model is still a cornerstone for our understanding of the process. A major puzzle remained, however. Asymmetric fission of the actinides became understandable not before the shell model became known. It was Maria Goeppert-Mayer, one of the co-discoverers of the model, who in 1948 pointed to the connection between fragment mass-asymmetry and closed neutron shells. Further major steps on the experimental side were the discovery of spontaneous fission of Uranium by G. N. Flerov and K. A. Petrzak in 1940 and of ternary fission in 1947 by Tsien San-Tsiang.

Traditionally nuclear fission was studied by radiochemical and physical methods. The physical techniques employed were photographic emulsions, ionization chambers, solid state and TOF detectors. As a new tool for fission studies, the idea came up to develop a mass separator for fission fragments based on magnetic and electric fields. In the early sixties of last century this idea took shape [1]. With the advent of powerful reactors and with fissile targets placed near the core, the count rates for fission fragments recoiling from the target and being intercepted by the mass separator became competitive. In view of the high resolution for masses and energies of fragments this was an intriguing new option. Pushed by P. Armbruster the mass separator "Lohengrin" was installed at the high flux reactor of the ILL and became fully operational in 1975 [2]. This was the start signal for fission studies at the ILL. For specific studies complementary detector systems based on ionization chambers and time-of-flight devices were built (Cosi fan tutte, Diogenes). A particular line of research was the investigation of binary and ternary fission induced by polarized neutrons. A further outstanding feature at the ILL is besides high neutron flux the generalized use of curved neutron guides which allow to take data under very clean background conditions e.g. for the measurement of fission cross sections.

It has to be stressed that all fission investigations to be presented in the following were performed for thermal and cold neutrons.

2. Instruments and Methods

The work horse for fission experiments at the ILL is the mass separator "Lohengrin". The layout of Lohengrin is shown in the left panel of Fig. 1. It has been in operation for more than 3 decades and continues to be in high demand. It serves as a separator for unslowed fission fragments recoiling from a thin target of fissile material placed inside the heavy water moderator of the reactor close to the core. It is a focussing parabola spectrograph with a magnetic sector field and an electric cylinder condenser. The fragments have to travel 23 m from the source to the focal plane corresponding to a travel time of about 2μs. Fission products (after prompt neutron evaporation) are separated along parabolas into spectra of A/q lines with A the mass and q the ionic charge of fragments. Along the parabolas there is a dispersion in kinetic energy E. The resolution of the instrument is outstanding: $A/\Delta A \approx 1500$ and $E/\Delta E \approx 1000$ (FWHM). For applications e.g. in spectroscopy of n-rich fragments the high E-resolution is not required and a focussing magnet near the focal plane allows to reduce the dispersion (RED magnet in Fig. 1) [3]. Finally, to identify nuclear charges Z of fragments, their Bragg curves are inspected by suitable combinations of ΔE and E_{rest} detectors.

An example of a detector combination designed for ternary fission studies is given in the right panel of Figure 1 [4]. There are two concentric ionisation chambers centred on a fissile target. The inner chamber stops the fission fragments while the outer chamber intercepts long range ternary particles. In addition to particle energies also azimuthal and polar angles are deduced event by event yielding rather complete information on ternary processes. Note that the scales in Figure 1 for the two instruments differ by a factor of about 20.

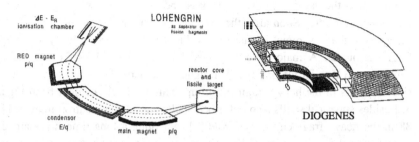

Figure 1. LOHENGRIN separator (left panel), DIOGENES (right panel) for ternary fission studies.

A major challenge has been the measurement of fragment charges. All techniques having been proposed exploit the fact that for given mass A and energy E of a fragment the ionisation curve still depends on the nuclear charge Z. Two variants are on display in Figure 2. To the left an ionisation chamber for use on LOHENGRIN with segmented anode (labelled 3 and 6) and a separation grid (labelled 8) is shown enabling to determine the energy loss ΔE and the rest energy E_{rest} along the ionisation curve [5]. In another approach the ranges of ions are measured in a Bragg ionisation chamber for mass and energy separated fragments as determined by a start-stop time-of-flight and an energy measurement (in the figure only the stop detector is drawn). From the range the charge Z is inferred [6].

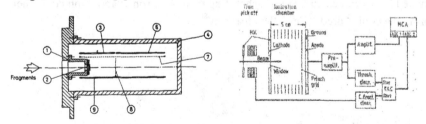

Figure 2. ΔE-E_{rest} ionisation chamber on LOHNEGRIN (left panel), Bragg ionisation chamber measuring ranges on Cosi fan tutte (right panel).

Unfortunately, all attempts to deduce nuclear charges Z by analyzing the shape of the Bragg curve for freely recoiling fragments has not allowed reaching charge resolutions much better than $Z/\Delta Z \approx 42$. It means that only for the light fragment group charge data are available while even for symmetric fission the charges remain unknown.

3. Mass Distributions of Fission Fragments

Mass distributions of fission fragments exhibit many interesting features. Most prominent is the asymmetric split of mass and charge in low energy fission of the lighter actinides. Soon after the discovery of the nuclear shell model it was recognized that asymmetric fission is due to the extra stability of magic fragment nuclei. A major role is played by the spherical shells N = 82 for neutrons and Z = 50 for protons. Throughout the actinides it fixes the onset of sizable yields in the heavy fragment group. This is clearly evident from Figure 3. Besides spherical shells also deformed shells e.g. for the neutron number N ≈ 88 in the heavy fragments play a role [7]. For quite a long time it remained a puzzle why shell effects in the light fragments were not showing up. The issue has only recently found an answer by experiments on LOHENGRIN where the mass and charge yield data were pushed to very light fragments. Results for 6 thermal neutron induced reactions have been collected in Figure 3 [8]. Besides the well known phenomenon of asymmetric fission traced to nuclei near the doubly magic ^{132}Sn, there is a second mass region, this time in the light fragments, where the yields become identical. In Figure 4 a zoom of Figure 3 is presented which allows for a more transparent interpretation. In the left panel of the figure yields around mass number A = 80 are depicted [9]. It is remarkable that for mass A = 80 there is a kink in the slope of the yields which is common to all reactions studied. It is suggested that behind mass A = 80 there is the magic neutron number N = 50. Isotopic mass distributions show that for this mass region the most probable charge is Z = 32 (Ge). The isotope ^{82}Ge would indeed be a good candidate for a shell fragment. Neutron evaporation could then bring the conjectured ^{82}Ge to the observed ^{80}Ge.

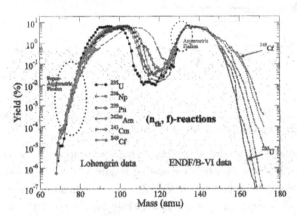

Figure 3. Asymmetric and super-asymmetric mass distributions in thermal neutron induced fission.

More spectacular than the kinks are shoulders in far-asymmetric fission revealed in the right panel of Figure 4 for a set of thermal neutron reactions [10]. The peaks of the shoulders are at mass number A = 70. They are not equally pronounced for all reactions. Looking for a possible structure effect in the fragments the stabilising influence of Ni-isotopes with the magic charge number Z = 28 comes to mind. That Ni-isotopes are accountable for the shoulders could directly be proven in experiment by measuring isotopic mass distributions [11, 12].

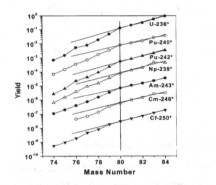 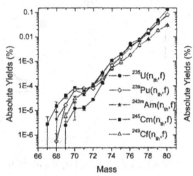

Figure 4. Fission yields around mass A = 80 (left panel; note that yields are displaced by a factor of ten for the sake of clarity); fission yields around mass A = 70 (right panel).

4. Charge Distributions of Fission Fragments

The investigation of fragment charge distributions has been for many years at the focus of research at the ILL, both on the instruments LOHENGRIN and COSI. Since with these instruments the time required for the determination of charge Z is much shorter than any intervening β-decay, the charges measured are those of the primary fragments. Conservation of charge therefore implies that the compound charge Z_{CN} is equal to the sum of the light and heavy fragment charges Z_L and Z_H, respectively. The limitations in charge resolution to charges Z < 45 is therefore not a major drawback for asymmetric fission but hampers the study of symmetric fission.

A sample of charge distributions is presented in Figure 8. The reactions on display are for the compound nuclei ^{230}Th* [13], ^{236}U* [14], ^{239}Np* [15], ^{242}Pu* [16], ^{246}Cm* [17] and ^{250}Cf* [18]. In the data a pronounced even-odd staggering for elements with even charges Z_{CN} shows up which fades away, however, for the heavier actinides studied. Systematically the yields for even fragment charges are larger than for odd charges. For the element Np with odd charge number Z_{CN} no staggering is observed in the range of charges shown in Figure 5.

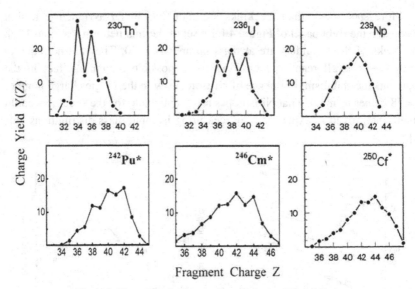

Figure 5. Charge distributions in thermal neutron induced fission.

The even-odd effect of charge yields has been extensively studied and discussed. Commonly the size δ_Z of the e-o effect is quantified by evaluating for even and odd fragment charges the yields Y_e and Y_o, respectively, with [19]

$$\delta_Z = \frac{Y_e - Y_o}{Y_e + Y_o} \quad . \qquad (4\text{-}1)$$

Usually the sum $(Y_e + Y_o)$ is normalized to 100 % and the charge e-o effect is quoted in %. A summary of all ILL results on charge e-o effects for even compound charges are visualized in the left panel of Figure 6 as a function of the compound mass number A. There is a general decrease of δ_Z with increasing mass. However, for the U- and the Pu-isotopes the effect stays constant within

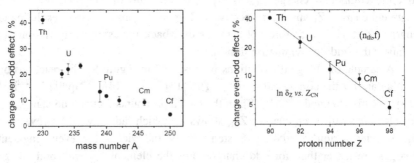

Figure 6. Charge e-o effect vs. compound mass A (left panel) and vs. compound charge (right panel).

error bars. It tells that for a given element the effect does not depend on neutron number. Taking averages for the U- and Pu-Isotopes the e-o effect is plotted as a function of compound charge Z_{CN} in the right panel of Figure 6. On a logarithmic scale the effect decreases linearly with Z_{CN}. The dependence is very smooth. In the past there were many discussions which parameter for the presentation of e-o effects should be best appropriate to bring a correlation into view which is physically motivated. At first the fissility parameter Z^2/A was chosen. The correlation being not really convincing the Coulomb parameter $Z^2/A^{1/3}$ was proposed as a better choice. However, the present parameterization with the charge Z_{CN} appears to be the best one. Yet, none of the parameters is really satisfactory since there is no hint for a more physical insight.

Trying to understand the dependence of the e-o effect on the fissioning nucleus a helpful correlation emerges when the effect is considered as a function of the total excitation energy TXE of the system. The excitation energy TXE is found from the difference between the Q-value of the reaction and the total kinetic energy TKE of fragments: TXE = Q – TKE. In figure 7 (left panel) TXE is plotted on a linear scale *versus* fissility Z^2/A of the compound nucleus. From experiment a linear relationship between the two quantities is observed with TXE increasing monotonously with fissility. This tendency is also borne out by theory. It just reflects the fact that from Th to Cf the scission point moves farer away from the saddle and hence the fragment deformations at scission become larger. Following scission the deformations relax and give rise to intrinsic excitations set free by prompt neutrons and gammas. The close correlation between average TXE and fissility translates with the help of the correlation between charge e-o effect and nuclear charge (or fissility) in Figure 6 (right panel) into the dependence of the e-o effect on TXE. This is shown in Figure 7 (right panel). On a logarithmic scale the e-o effect decreases linearly for increasing excitation energy TXE. It gives from experiment a direct indication that the charge e-o effect is connected to energy partition in fission.

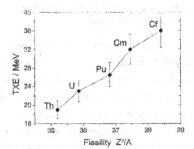 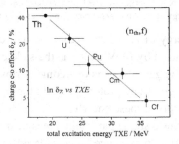

Figure 7. Total average excitation energy TXE *versus* fissility Z^2/A (left panel); charge e-o effect *versus* total excitation energy TXE for thermal neutron fission of even-Z nuclei.

Probably even more telling is the dependence of the charge e-o effect on the excitation energy E_{SAD}^* at the saddle point. In Figure 8 (left panel) the e-o effect δ_Z has been plotted for thermal neutron and MeV neutron induced fission of several U-isotopes as a function of E_{SAD}^*. To lump together different U-isotopes on a common footing is justified on the basis of Figure 6 (left panel) where δ_Z for the U-isotopes is seen to be identical for thermal neutron energies [20]. The labels U3 through U9 stand for $^{232}U(n_{th},f)$ through $^{238}U(n_{3MeV},f)$. The excitation energy at the saddle is found from

$$E_{SAD}^* = B_n - V_{BAR} + E_n \quad (4-2)$$

with B_n the neutron binding energy, V_{BAR} the barrier height and E_n the energy of the incoming neutron. For thermal neutron fission of fissile isotopes the saddle energy E_{SAD}^* is

$$E_{SAD}^* = (B_n - V_{BAR}) < 2\Delta \quad (4-3)$$

and hence below the pairing gap a the saddle $2\Delta \approx 1.7$ MeV. Only collective transition channels are excited. Remarkably, in Figure 8 up to an excitation

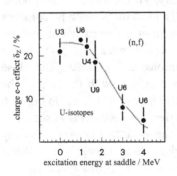 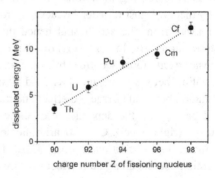

Figure 8. Charge e-o effect *versus* excitation energy at saddle for U-isotopes (left panel); dissipated energy at scission *versus* charge number Z_{CN} of compound nucleus (right panel).

energy of about 2Δ above the barrier the e-o effect stays constant. For fission induced by MeV neutrons the saddle energy will become $E_{SAD}^* > 2\Delta$. As soon as the gap energy is overcome intrinsic qp-excitations are populated by breaking nucleon (here proton) pairs. In the figure the e-o effect is observed to drop down sharply at these energies. This demonstrates that the e-o effect is sensitive to the intrinsic single particle degrees of freedom but not to the collective ones. The e-o effect may hence be considered as a sensor for the intrinsic excitation energy E_X^{sci} already present at scission. This is a valuable feature because it allows to disentangle the contributions to the total final excitation energy TXE of E_X^{sci} and the energy V_{Def} of deformation at scission:

$$TXE = E_X^{sci} + V_{Def}. \qquad (4\text{-}4)$$

In Figure 8 the drop of the e-o effect is due to thermal excitation of single particle states in the compound stage of the fissioning nucleus. More generally pair breaking may occur in the course of fission from saddle to scission. Where and by which mechanism pairs are broken in the elongation and scission process is still open to discussion. For the (e,e)-nuclei considered one may think of two limiting cases: either the flow of nuclear matter is viscous leading to thermal excitations, or the flow is superfluid with pairs being broken by non-adiabatic processes very close to the instant of scission. Non-adiabatic processes near scission are expected because at scission a potential barrier between the two nascent fragments rises like a volcano and the snapping of the neck is very fast.

Evidently the question is, how can the excitation E_X^{sci} at scission be assessed quantitatively from a measurement of the charge e-o effect δ_Z. The discussion of this question is controversial. There are, first statistical models where it is assumed that the intrinsic degrees of freedom reach thermal equilibrium before scission is reached. In a second class of models an equilibrium is not postulated. It is pointed out that due to the weak coupling between collective and single particle degrees of freedom e-e nuclei could remain fully paired up to the loss of adiabaticity at scission. Pairs are then broken in the very process of scission.

One of the first models proposed is an innovative combinatorial analysis of the probability distribution of pairbreaking where the mechanism of breaking is left open [21]. Starting from a superfluid model for the nucleus, pairbreaking is brought about by qp-excitations. With the definitions for

N = maximum number of 2 qp-excitations depending on excitation energy being available,

q = probability to break a pair when the energy is available,

ε = probability for the broken pair to be a proton pair, and

p = probability for nucleons from a broken pair to go into complementary fragments, the charge e-o effect δ_Z is calculated to be

$$\delta_Z = (1 - 2p\varepsilon q)^N. \qquad (4\text{-}5)$$

The dissipated energy, on the other hand, is given by the average number <N> of broken pairs and the energy 2Δ required to break a pair. With <N> = qN it therefore follows

$$E_X^{sci} = 2\Delta \cdot <N> = 2\Delta\, qN. \qquad (4\text{-}6)$$

The only parameter known with some certainty in these equations is the gap parameter which at the saddle point is $2\Delta \approx 1.7$ MeV and at the scission point $2\Delta \approx 2.4$ MeV [22]. It is further plausible that for ε the relative probability for proton breaking is just $\varepsilon = Z/A$ with Z and A the charge and mass number of the fissioning nucleus. Therefore $\varepsilon \approx 0.4$ should be a good approximation for the nuclei under study. For a proton pair broken in the descent from saddle to scission the two protons may be conjectured to stay either together in one fragment or to go into complementary fragments. This would give p = 0.5. However, for pairbreaking near neck rupture p is expected to be more close to p ≈ 1. The size of the remaining parameter q remains unknown. Assuming q = 0.5 as a first guess yields with $2\Delta \approx 1.7$ MeV and with p ≈ 0.5 roughly

$$E_X^{sci} \approx -3.8 \ln \delta_Z \text{ MeV.} \tag{4-7a}$$

For pairs being broken at scission one finds with $2\Delta \approx 2.4$ MeV and p ≈ 1

$$E_X^{sci} \approx -2.4 \ln \delta_Z \text{ MeV} \tag{4-7b}$$

The formula tells that E_X^{sci} is nil for no pair-breaking (N = 0, δ_Z = 1), while E_X^{sci} increases the more pairs are broken (N > 0, $\delta_Z \to 0$). Scission energies calculated with p ≈ 0.5 (eq. 4-7a) for thermal neutron fission from Th to Cf are displayed in Figure 8 (right panel) *versus* the charge number of the fissioning nucleus. The energies range from $E_X^{sci} \approx 3.5$ MeV for Th up to $E_X^{sci} \approx 12.3$ MeV for Cf. With the choice p ≈ 1 the excitation energies at saddle are smaller by a factor of 1.6. The uncertainty in the parameters p and q and hence in the evaluated energy E_X^{SCI} is sizable but in both approaches the intrinsic excitation energies at scission are inferred to be small. They should be compared to the total excitation energies TXE in Figure 7 (left panel). The excitation at scission contributes from 20% to 30% to the total excitation energy.

So far the e-o staggering of the global charge distribution was considered. More in detail one may wonder whether the size of the staggering does snot depend on the charge Z itself. To this purpose a local e-o effect has been proposed assessing the effect for a limited range of 4 consecutive charge numbers as the deviation from a smooth Gaussian at the average <Z> = Z + 1.5 [23]:

$$\delta_Z(<Z>) = \tfrac{1}{8}(-1)^{Z+1}[\{\ln Y(Z+3) - \ln Y(Z)\} - 3\{\ln Y(Z+2) - \ln Y(Z+1)\}]. \tag{4-8}$$

Two examples of local e-o effects are on display in Figure 9. To the left the local effect is seen to rise sharply for super-asymmetric fission of ^{236}U* [24].

Even more surprising is the pronounced e-o effect for super-asymmetric fission of the odd-Z_{CN} nucleus Neptunium (^{239}Np*; right panel) [25]. In the global distribution of charge for ^{239}Np* in Figure 5 no e-o staggering could be discerned. This seemed to be perfectly compatible with the presence of an unpaired proton in Np which was thought to be free to join either of the two fragments. However, in super-asymmetric fission the proton apparently goes preferentially to the heavy fragment leaving the lighter fragment shown in the figure in a paired configuration. This behaviour has been well explained in a statistical model [22]. It is pointed out that the sticking probability is proportional to the single particle level density in fragment Z which is proportional to Z / Z_{CN}. In asymmetric fission the sticking probability to the heavier fragment is hence larger than for the light fragment.

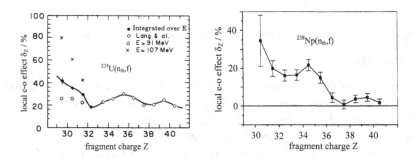

Figure 9. Local e-o effect for ^{236}U* (left panel) and 239Np* (right panel).

In Figure 9 a marked dependence of the e-o effect on the kinetic energy of fragments is visible for fission of ^{236}U*. The dependence of the e-o staggering on the kinetic energy of fragments is a general phenomenon. This is shown in Figure 10 for the two fissioning Pu-isotopes ^{240}Pu* [26] and ^{242}Pu* [27]. The global e-o effect is plotted as a function of the light fragment kinetic energy E_{KL}. The effect is large for high kinetic energies E_{KL} and decreases for decreasing energy. Since with E_{KL} also the total kinetic energy TKE as the sum of the two fragment energies decreases and, since from energy conservation the available energy Q = TKE + TXE has to be shared, a decrease in TKE is equivalent to an increase of the total excitation energy TXE. It was already noted in Figure 7 (right panel) that for an increase in TXE the e-o effect δ_Z slopes down. Therefore the present observation is not a surprise. It just confirms the previous discussion following eq. (4-7) that the excitation at scission E_X^{sci} controlling the e-o effect is a fraction from 20 % to 30 % of TXE and hence the increased E_X^{sci} brings the e-o effect down.

However, these arguments describe the situation only for light fragment kinetic energies larger than about 90 MeV. For lower kinetic energies the e-o effect stays constant or even increases again. This feature is surprising and will be analyzed in connection with the phenomenon of cold fission in the next paragraph 5.

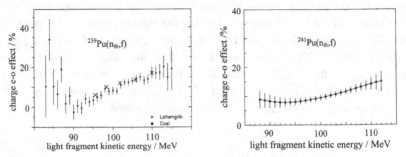

Figure 10. Charge e-o effect as a function of light fragment kinetic energy for 240Pu* and 242Pu*.

5. Cold Fission

From studies of thermal neutron induced fission it was already in 1961 reported that for a broad fragment mass range the total kinetic energies TKE may exhaust all of the available energy Q [28]. The discovery became known as "cold fission". Dedicated experiments investigating cold fission started in 1981 [29, 30]. In principle the LOHENGRIN spectrometer is well suited to search for rare events and cold fission is a rare process. To reach cold fission the kinetic energies have to be pushed as high as possible. A mass distribution for ^{235}U(n,f) taken at a rather high kinetic energy of the light fragment E_{KL} and compared to the global yield Y(A) in Figure 11 (left panel) demonstrates vividly that the mass distribution changes dramatically [31]. At this energy one is near n-less fission and this enabled identifying two of the mass peaks with magic nuclei.

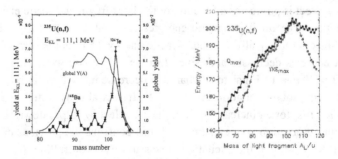

Figure 11. Mass distribution for ^{236}U* at high E_{KL} = 111.1MeV (left panel); comparing TKE$_{max}$ to Q$_{max}$ for ^{236}U* delineating the mass region with cold fission (right panel).

For still higher energies the mass distributions become fragmented in the sense that no longer all masses can contribute. For a discussion of cold fission this is not convenient. An alternative presentation of data is provided from measurements with a twin back-to-back ionisation chamber. The two chambers share a common cathode which at the same time serves as the backing for fissile material. When irradiated by thermal neutrons the two complementary fragments are intercepted by the two chambers. Besides fragment masses the total kinetic energy TKE is found. An example for the standard ^{235}U(n,f) reaction in Figure 11 (right panel) brings directly to evidence that the maximum TKE$_{max}$ is nearly coinciding with the maximum Q-value Q_{max} for a broad mass window from $A_L \approx 85$ to $A_L \approx 106$ in the light fragment group [32]. Needless to say that neutron emission is excluded for this mass range. It has further to be pointed out that for symmetric and very asymmetric mass splits cold fission is far from being attained. For symmetric fission this could be anticipated from the known kinetic energy dip in this region of mass. In a model calculation searching the most compact scission configurations corresponding to true cold fission these properties were well reproduced [33].

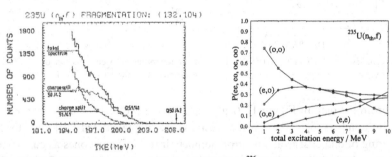

Figure 12. High TKE tail for the fragmentation (132,104) in ^{236}U* (left panel); breakdown of yields for (Z,N) fragments as a function of (Z,N) dependent total excitation energy (right panel).

For a more precise analysis it is, however, necessary to identify not only fragment masses but also their nuclear charges Z. It means that Q-values for charges have to be taken into account and not merely the charge maximising the Q-value. This is a demanding task. As an example, for the mass fragmentation (132,104) of ^{236}U* the events in the high TKE tail have been sorted according to the charge ratio Z_H/Z_L in Figure 12 (left panel) [34]. Unexpectedly it is not at all the charge ratio 50/42 with the maximum Q-value for which "true" cold fission is realized. The goal is missed by some 3 MeV which will appear in the total excitation energy TXE. Instead, events with the o-o charge ratio 51/41 virtually reach the respective Q-value. By inspecting all mass fragmentations the relative yields of fragments with (Z,N) equal to (o,o), (e,o), (o,e) and (e,e) are found as a function of the excitation energy TXE. The result is plotted in Figure 12 (right panel) [34]. True cold fission with the total TXE < 1 MeV (the experimental uncertainty as to absolute values) is only populated by (o,o) fragments. Odd mass fragments stay away by about 1 MeV and (e,e) fragments by about 2 MeV from true cold fission. These findings remind the systematic of level densities. In the back-shifted Fermi gas model effective excitation energies U^* are introduced. For (o,o) nuclei one has $U^* = U$ with U the physical excitation. For o-mass nuclei $U^* = U - \Delta$ and for (e,e) nuclei $U^* = U - 2\Delta$. With an average gap parameter of $\Delta = 1$ MeV for fission fragments it is seen that the fission yield is setting in for the effective $U^* \geq 0$ MeV. The level density is thus the dominant factor controlling the outcome of cold fission. Very similar results were obtained on LOHENGRIN though with somewhat larger systematic uncertainties [35].

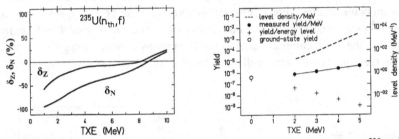

Figure 13. Even-odd effect for charge and neutron numbers close to cold fission of ^{236}U* (left panel); yields per energy level extrapolated to the ground state for ^{82}Ge + ^{152}Nd for ^{234}U* (right panel).

From the data in Figure 12 (right panel) the e-o effect for both charge and neutron numbers, δ_Z and δ_N, respectively, is readily calculated [36]. Due to the favouring of odd charge and neutron numbers instead of even ones as customary from standard fission, the e-o effect is negative close to cold fission. The reversal of signs for the e-o effects was not anticipated and up to now has not been scrutinized by theory. While level densities are decisive for the observation of cold fission in experiment it is open to discussion whether instead of level density the probability of populating ground states would not be a better criterion. This would directly address the matrix element for the transition. Unfortunately the feeding of ground states could not be observed in experiment. Yet, an attempt was made to extrapolate yields, level densities and the yield per level determined in the excitation range TXE from 2 MeV to 5 MeV for ^{234}U* to find the probability for feeding ground state yields [35]. This is indicated in Figure 13 (right panel). Unfortunately data were only taken for a limited mass range from $A_L = 76$ to $A_L = 93$. It turns out that light fragments with even charges Z_L have in most cases higher yields than odd charges and among those (e,e) fragmentations are favoured compared to (e,o) splits. For example for (e,e) fragments the ground state yields per level are about 10^{-7} per fission. Because of uncertainties in the energies of fragments the absolute sizes of the yields deduced may have sizable error margins, but the above systematic of relative yields should be independent from this shortcoming. It has hence to be concluded that, with the probability for feeding ground states as the criterion, the e-o effect is positive enhancing the yields of even charges compared to odd charges.

In the above the emphasis was put on charge distributions because these are those of the primary fragments. By contrast, the neutron numbers of fragments determined in experiment are usually those after prompt neutron emission. Since neutron evaporation itself depends on the (e,o) character of nuclei, any genuine primary e-o effect is masked. The difficulty does not arise for neutron-less fission. The probability for neutron-less fission is 3.2 % for thermal neutron fission of ^{236}U* and only 0.23 % for ^{252}Cf(sf). But one should not confound n-less fission with cold fission. True cold fission with excitation energies not exceeding TXE ≈ 1MeV is only a tiny fraction of n-less fission which is observed for TXE up to 10 MeV if not higher up.

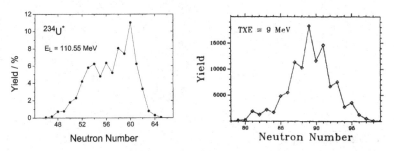

Figure 14. Neutron distribution for ^{234}U* at the light fragment energy E_{KL} = 110.55 MeV (left panel); neutron distribution at the excitation energy TXE = 9 MeV (right panel).

Two examples for neutron e-o effects are presented in Figure 14. To the left the neutron distribution for ^{234}U* at the light fragment energy E_{KL} = 110.55 MeV reveals an e-o staggering favouring even neutron numbers [37]. At this energy the process is not really n-less but very close to it and therefore neutron emission should be strongly suppressed. In the right panel the neutron distribution for n-less fission in ^{252}Cf(sf) at a total excitation energy of TXE = 9 MeV is shown [38]. The two figures convey different physics. For fission of ^{252}Cf the e-o effect has changed sign. Now the odd neutron numbers carry the larger yield. With no theory having been proposed the phenomenon will be discussed in simple terms.

To introduce this discussion it will be useful to recall an often quoted model, the scission point model. It is visualized in Figure 15 (left panel) [39]. For a scission configuration consisting of two fragments facing each other while their deformations increase, the potential Coulomb and deformation energies V_{Coul} and V_{Def}, respectively, are plotted together with their sum ($V_{Coul} + V_{Def}$). The energy relations are summarized for convenience and read

$$Q = TKE + TXE \qquad (5\text{-}1)$$
$$TKE = V_{Coul} + E_K^{sci} \qquad (5\text{-}2)$$
$$TXE = V_{Def} + E_X^{sci}. \qquad (5\text{-}3)$$

The difference between the total available energy Q and the sum of potential energies is labelled E_{Free}. E_{Free} is the sum of energies at scission not bound as potential. Its range is marked by the hatched region in the figure:

$$E_{Free} = E_K^{sci} + E_X^{sci}. \qquad (5\text{-}4)$$

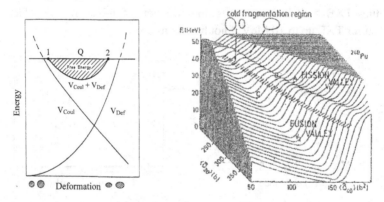

Figure 15. Energies *vs.* deformation of fragments at scission (left panel); microscopic calculation of potential energy of deformation from saddle to scission for ^{240}Pu*.

Two limiting cases appear labelled as 1 and 2 in the figure. They correspond to the most compact and the most deformed scission configuration with maximum and minimum total kinetic energy TKE, respectively. In these two cases $E_{Free} = 0$. Vanishing E_{Free} entails $E_K^{sci} = E_X^{sci} = 0$. In general one has $E_{Free} \neq 0$ but from experiment it is unfortunately not known how the energy is distributed. One has to invoke theoretical models. The question was already addressed when discussing charge e-o effects. For the sake of simplicity let us assume either fission is viscous with $E_K^{sci} \equiv 0$ or fission is adiabatic with $E_X^{sci} \equiv 0$. Of course, intermediate situations can not be ruled out.

Along the lines of the scission point model it is convenient to come back to a discussion of Figure 11 where the charge e-o effect was displayed as a function of fragment kinetic energy. From high towards lower kinetic energies the e-o effect decreases. In the scission point model one moves thereby from compact to more deformed configurations. The free energy rises and, according to the two approaches proposed, either this tells that E_X^{sci} or that E_K^{sci} increases. This leads to the interpretation that proton pairs are broken either due to higher thermal excitation or due to the more efficient loss of adiabaticity being imposed by the higher speed of receding fragments at scission. The thermal excitation arguments were put forward in connection with Figure 10. The re-enforced e-o effect at very low kinetic energies in Figure 10 is, however, better understood in the adiabatic model. It pertains to those events having not undergone fission near the minimum of the free energy in Figure 15 but where the speed of fragments is slowed down when they have to climb up the potential towards cold deformed fission at point 2. For reduced speeds the probability of pair-breaking is lower and the e-o effect comes back.

Apart from fragmentation theory a very successful model of cold fission has been proposed in the framework of constrained Hartree-Fock calculations [40]. The potential energy surface for the fissioning nucleus ^{240}Pu is on display in Figure 15 as a function of the multipole moments Q_{20} and Q_{40} (right panel). Nuclear matter is taken to be non-viscous. In the figure the fission valley for the mono-nucleus starts at the saddle and slopes down towards scission. The valley is separated from the fusion valley with two separated fragments by a ridge. The height of the ridge gets smaller and eventually disappears on approach to the scission point. However, before reaching this point, the collective stretching motion may be coupled to a transversal collective mode of necking-in. Either by tunnelling through the ridge or, after having acquired sufficient energy by passing over the ridge, fission into the fusion valley may occur for only slightly deformed nuclei. Cold compact fission thus finds a convincing interpretation with predicted rates comparable to experiment. It has to be stressed that in this theory pair breaking leading to e-o effects occurs in the process of violent scission. It should also be noted that the phenomenological scission point model is not in conflict with this solid physical theory.

With a look at the PES of the microscopic model in Figure 15 it is tempting to venture a speculation for the interpretation of negative e-o effect in cold fission. To simplify the discussion it is assumed that at most one pair of nucleons can be broken. With the normalization either for protons or neutrons of the even and odd yields $1 = (Y_e + Y_o)$ one finds for the e-o effect from eq. (4-1)

$$\delta = (Y_e - Y_o) = (1 - 2Y_o). \tag{5-5}$$

The yield Y_o for a fragment with an odd proton or neutron number emerging from the breaking of a nucleon pair is equal to the probability q that indeed a pair has been broken, times the probability p that the two quasi-particles from the pair end up in different fragments. Therefore

$$\delta = (1 - 2pq). \qquad (5\text{-}6)$$

For pair-breaking due to the necking-in volcano, which is splitting the mother nucleus into two halves, it is reasonable to assume that the two partners of the pair are separated and go to complementary fragments. It means that throughout the probability p may be set to p = 1. The only remaining crucial parameter is the probability q of pair-breaking. In true cold fission (TXE < 1 MeV) with only (o,o) fragments present (see Figure 12) necessarily both a proton and a neutron are broken for sure and hence q = 1. With eq. (5-6) one finds a maximum negative even-odd effect for both, protons and neutrons with $\delta_Z = \delta_N = -100\%$. This is shown in the left panel of Figure 13. The reason why the e-o staggering becomes less negative and eventually positive for increasing excitation energy TXE may find an intuitive answer by noting that for increasing TXE the deformations of fragments at scission increase. By the same token the thickness of the neck to be cut at scission which is rather thick in cold fission will become thinner the larger the excitation energies are. Due to the short range of the pairing interaction it is therefore to be expected that less nucleon pairs will be affected by the scission process in standard fission than in cold fission. Hence, when going from cold to standard fission the probability q for breaking will decrease from q = 1 to values q < 1 and in consequence the even-odd effect in eq. (5-6) will rise towards $\delta = 0$ and eventually turn from negative to positive.

It has to be kept in mind that the suggested influence of the neck thickness is only one of the parameters controlling pair-breaking. In the adiabatic model another parameter is the velocity of the fragments at scission which in turn governs the speed of neck closure and therefore the loss of adiabaticity. There is no excitation energy E_X^{sci} at scission, instead energy from the necking-in mode is dissipated and goes into the energy required for breaking nucleon pairs.

6. Ternary Fission

In ternary fission besides the two main fission fragments a third charged particle is ejected. By far in most cases the ternary particles are α-particles with ranges e.g. in photonuclear emulsions much longer than the tracks of fission fragments. But there are many other species of ternary particles showing up. The

LOHENGRIN spectrometer has been an ideal tool to investigate systematically the yields and energy spectra of these particles down to very low yields.

A systematic of ternary yields per fission is on display in Figure 16 for thermal neutron reactions from ^{236}U* to ^{250}Cf* (left panel). Yields down to 10^{-10} / fission were determined. It is observed that the heavier the fissioning nucleus is, the heavier are also the ternary particles which are ejected. For ^{234}U* the heaviest isotopes detected were ^{24}Ne and ^{27}Na. For the reaction ^{249}Cf(n$_{th}$,f) ternary silicon and sulphur nuclei are found with mass numbers up to almost 40 with yields at 10^{-8} per fission. Still heavier elements are to be expected when pushing the measurements to yields of 10^{-10} per fission like for 234U*. Generally there is a pronounced even-odd effect in the yields with elements with an even charge number being produced with higher probability.

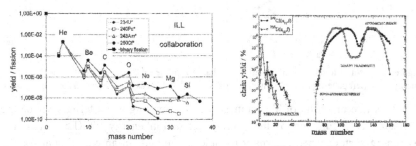

Figure 16. Systematic of ternary yields / fission for the reactions ^{233}U(n,f) to ^{249}Cf(n,f) (left panel); synopsis of yields for binary fragments and ternary particles in ^{236}U* and ^{250}Cf* fission (right panel).

In a synopsis of binary fragment and ternary particle yields in Figure 16 (right panel; normalization to 100 fission events) it is tempting to imagine a correlation between super-asymmetric binary fission and the heaviest ternary yields. For fission events with the cluster ^{132}Sn in the heavy fragment group and the cluster ^{70}Ni in the light group (see Figure 4) the number of "free" nucleons is larger for ^{250}Cf* than for ^{236}U*. This would allow a heavier ternary particle to be emitted from the heavier Cf nucleus. For the time being this is just a conjecture which should not be squeezed to quantitative comparisons on the basis of the presently available data.

Details of yield measurements for the reaction ^{242}Am(n,f) are summarized in Figure 17 (left panel). For the sake of clarity only yields for particles with even Z are visualized. Data for odd particle charges are similarly available [41]. The yields are plotted as a function of the "energy costs". Energy costs are introduced as the crucial parameter in a basic theory of ternary fission yields [42]. The yields are proposed to be given by

$$Y \sim \exp(-E_{cost}/T) \quad \text{with} \quad E_{cost} = E_{bind}^{TP} + \Delta V_C^{ter-bin} \qquad (6\text{-}1)$$

where E_{bind}^{TP} is the binding energy of the ternary particle to a fragment and $\Delta V_C^{ter-bin}$ is the Coulomb energy which is required to place a ternary particle between the two main fragments. The correlation in Figure 17 reproduces the trends of measured yields satisfactorily well. It is tempting to identify the fit parameter T in eq. (6-1) as a temperature but in the spirit of the model the energy costs are drained form the deformation energy at scission and T is a parameter ruling the probability for the energy transfer.

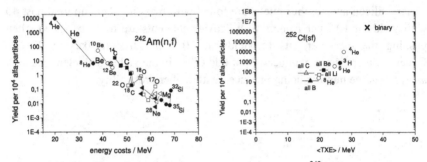

Figure 17. Yield of ternary particles with even charge Z in the reaction ^{242}Am(n,f) as a function of energy costs (left panel); yields of ternary particles from ^{252}Cf(sf) as a function of TXE (right panel).

This interpretation of T becomes more compelling when ternary yields are evaluated as a function of the average excitation energy <TXE> left to the fissioning system when a ternary particle is emitted. In an experiment for the reaction ^{252}Cf(sf) the energies of all three outgoing particles could be taken in coincidence [43]. Therefore the <TXE> could be determined for fission with well defined ternary particles. The result is on display in Figure 17 (right panel). From binary to ternary fission the excitation energy <TXE> remaining decreases the heavier the ternary particles are. Stated otherwise, the more TXE is drained the smaller are the yields of ternaries. Since the lion's share of TXE is tied up at scission as deformation energy it is hence recognized that the deformation energy supplies the energy for ternary particle emission.

Particularly for ternary long range α-particles a phenomenological model is discussed where the ratio LRA / B is described by

$$LRA/B = S_\alpha \cdot P(TXE > E_{cost}) \cdot P_{trans} \qquad (6\text{-}2)$$

with S_α the spectroscopic factor of α-preformation in the mother nucleus, $P(TXE > E_{cost})$ the probability that TXE at scission is larger than the required energy E_{cost}, and P_{trans} the probability that the TXE will be transferred to the ejection of the α-particle. The spectroscopic factor is extracted from the analysis of α-decay. The transfer probability is calculated in the sudden approximation. The ansatz (6-2) is in line with the above more general conclusions.

The importance of the spectroscopic factor S_α could be demonstrated by experiment. Results from a systematic of LRA/B ratios for a range of nuclei with varying fissility Z^2/A are shown for spontaneous and neutron induced fission in Figure 18 [45]. Yield ratios are plotted as a function of fissility to the left while to the right the same ratios but divided by the spectroscopic factor are displayed. Evidently the introduction of the spectroscopic factor significantly reduces the fluctuations in the experimental yields. Unfortunately however, except for the α-particle, spectroscopic factors for all other ternaries are not known from experiment. This restricts the application of spectroscopic factors in the assessment of yields to the ternary reaction at hand. As a side remark, it is seen in Figure 18 that in the spontaneous process the ternary α-yields are generally slightly enhanced.

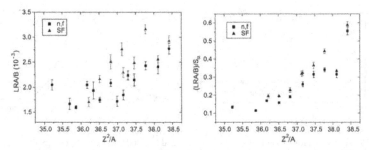

Figure 18. Yields of α-particles normalized to 10^3 binary events *versus* fissility for ^{235}U(n,f) and ^{252}Cf(sf) (left panel); same data as to the left but divided by the spectroscopic factor S_α (right panel).

Extending the analysis of ternary yields to all observable He-isotopes a so far not considered dependence for the probability of ternary fission comes into view. Yields of He-isotopes for the reactions ^{235}U(n,f) and ^{252}Cf(sf) are pictured in Figure 19 [46] (left panel). On a logarithmic scale the decrease in yield from 4He to ^{8}He is about linear. In particular for ^{252}Cf(sf), ternary yields were taken in coincidence with neutrons from the neutron unstable isotopes ^{5}He and ^{7}He decaying in flight. In experiment these isotopes are measured as ^{4}He and ^{6}He, respectively. The yields for the unstable ^{5}He and ^{7}He are usually underestimated. A spin weighting factor (2I + 1) for the odd-mass isotopes has to be introduced in order to bring experiment and model into agreement. This is a further parameter which should be taken into account. By contrast, the low yields for the isotope ^{3}He have to be underlined. In fact, ^{3}He has never been reported to be detected as a ternary particle, only upper limits could be determined [47]. For ^{235}U(n,f) the upper limit for the yield of ^{3}He is by a factor of 10^6 below the ^{4}He yield. All models of ternary yields have difficulties to cope with this large factor.

The yield of primary ^{5}He is only a factor of 5 smaller than the yield of the dominant ^{4}He whose LRA/B ratio is indicated in Figure 19. Due to its short life time of $T_{1/2} = 1.1 \cdot 10^{-21}$ s it is the ^{4}He from the decay ^{5}He $\rightarrow$ ^{4}He + n which is measured in the energy spectrum of alphas. For ^{252}Cf(sf) the residual ^{4}He from ^{5}He decay has an average kinetic energy of 12.4 MeV and is thus by 4 MeV lower than the average energy of 16.4 MeV of the primary ^{4}He. In consequence the experimental energy spectrum is a convolution of two energy distributions. This becomes visible as a low energy tailing. For the thermal neutron reaction ^{235}U(n,f) the experimental spectrum and its deconvolution into two components is on display in Figure 19 (right panel) [48].

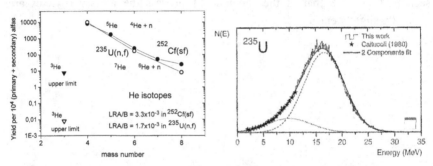

Figure 19. Ternary yields of He-isotopes in ^{235}U(n,f) and ^{252}Cf(sf) (left panel); energy distribution of ^{4}He deconvoluted into primary ^{4}He and ^{5}He (right panel).

Very briefly it should be mentioned that besides o-mass He-isotopes being unstable due to neutron decay also ternaries unstable with respect to charged particle decay have been discovered, both in neutron induced fission of U-isotopes and in spontaneous fission of ^{252}Cf [49]. The ternary particles and reactions in question are ^{8}Be $\rightarrow$ ($\alpha+\alpha$) and ^{7}Li* $\rightarrow$ ($\alpha+t$). While these reactions are "pseudo" quaternary fission events, in parallel "true" quaternary fission was observed with two heavy fission fragments and two charged lighter particles. Two examples of quaternary fission could be distinguished in thermal neutron fission of ^{234}U* and ^{236}U*. These are ($\alpha+\alpha$) with a yield per fission of $4 \cdot 10^{-8}$ and ($\alpha+t$) with yield $6 \cdot 10^{-9}$. For ^{252}Cf(sf) the quaternary yields are larger by about a factor of 10.

7. Fission Induced by Polarized Neutrons

High fluxes of polarized cold neutrons are available at the ILL on neutron guides. Typically the fluxes are ~ 10^9 n/s·cm² with energy 4 meV and polarization up to 95 %. These outstanding features make the study of very rare processes feasible. Of prime interest is the investigation of parity non-conservation (PNC) in nuclear fission which requires the use of polarized neutrons. The effect was studied in the past for many nuclei ranging from ^{229}Th up to ^{249}Cf. At the ILL the uniquely high flux enabled the comparison of PNC effects for binary and ternary fission. The best analyzed reaction with cold neutrons so far is ^{233}U(n,f).

The basic layout of the experiment is sketched in Figure 20. The target is at the centre of a reaction chamber which is traversed by the polarized neutron beam. Two multi-wire-proportional-chambers and two arrays of PIN diodes for detecting fragments and ternary particles, respectively, are centred on a plane perpendicular to the beam. The neutron beam is polarized with polarization facing the fragment detectors. The spin is flipped every second and data for the two spin orientations are compared with each other. In case parity is violated in fission the angular distribution of fragments (by convention the light fragment LF) relative to neutron spin is

$$W(\mathbf{p}_{LF}) \, d\Omega_{LF} \sim [1 + \alpha^{PNC}(\sigma_n \cdot \mathbf{p}_{LF})] \, d\Omega_{LF}. \tag{7-1}$$

Thereby α^{PNC} measures the size of the PNC effect, σ_n is the unit vector of spin orientation and $\mathbf{p}_{LF}$ is the unit vector of light fragment momentum. The pseudoscalar $(\sigma_n \cdot \mathbf{p}_{LF}) = \cos \Theta$ is a signature for parity violation. The PNC coefficient was measured for binary and ternary fission by requiring in the latter case the presence of a ternary particle. For the reaction ^{233}U(n,f) the results are

binary fission: $< \alpha^{PNC} > = [+0.396(17)] \cdot 10^{-3}$
ternary fission: $< \alpha^{PNC} > = [+0.37(10)] \cdot 10^{-3}$.

The size of the effect is as predicted from theory. What is new is the fact that for the two types of reactions the PNC effect is equal within error bars [50]. Since, as for regular angular distributions, also the PNC distribution is settled at the saddle point, the identical result reveals that binary and ternary fission share the same saddle point configurations. Ternary fission is thus a last minute effect decided only close to scission. This confirms ideas on ternary fission being the basic concept in most models.

With the same experimental setup but with the spin oriented along or against the beam two new phenomena were discovered in ternary fission. The experiment is sketched to the right in Figure 20. First, a so-called TRI effect was found [51]. Depending on spin orientation the yield for ternary particles emission is asymmetric with respect to a plane $(\sigma_n, \mathbf{p}_{LF})$. This is a curious effect whose interpretation is still at issue. Most probably it is the Coriolis force acting on the outgoing ternary particle which brings about an asymmetry. Second, the so-called ROT effect describes the observation that the ternary particles are set free by a system which is rotating. This induces shifts in the angular distribution of ternary particles. The shift angles are much smaller than 1° but could nevertheless be safely measured thanks to the spin flip technique [52, 53]. The effect proves that the fissioning compound undergoes a collective rotation.

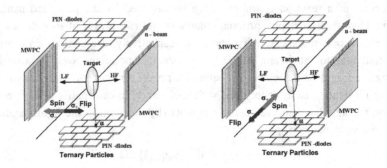

Figure 20. Sketch of the experimental layout for PNC investigations (left panel); the same but for investigations of TRI and ROT effects (right panel).

8. Fission Cross Sections

The excellent neutron beam conditions at the curved neutron guides of the ILL are well suited for the investigation of small fission cross sections in the sub-barrier energy range. A few examples obtained by the Geel group are presented.

As known since the early days of fission research, it is the pairing energy of the incoming neutron which for U-isotopes with an odd neutron number leads to an excitation energy of the compound which is sufficient to overcome the fission barrier. By contrast, for U-isotopes with an even neutron number fission following neutron capture proceeds as a sub-barrier process. The fission cross sections are hence small and their determination is a challenge. For the target nucleus ^{234}U the thermal neutron cross sections was found to be as small as (300±20) mb [54]. This figure has to be compared with the cross section e.g. of the neighbouring ^{235}U target which amounts to 584.3 b.

A similar jump of fission cross sections was reported for some Cm-isotopes [55]. For the target ^{243}Cm the thermal fission cross section is (572±25) b while for ^{246}Cm and ^{248}Cm the cross sections are (12±25) mb and (316±43) mb.

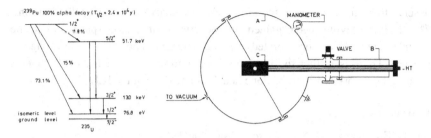

Figure 21. Decay scheme of ^{239}Pu (left panel); appliance for the production of ^{235}U^m (right panel).

An exceptional cross section measurement was performed for the isomeric state of ^{235}U. The ^{235}U^m state with spin $I^\pi = \frac{1}{2}^+$ has a halflife of $T_{1/2} = 26$ min. The ^{235}U^m target for cross section measurements has therefore to be produced in the immediate vicinity of the neutron beam. The isomer is formed in the α-decay of ^{239}Pu. As shown to the left in Figure 21 all decays lead in times shorter than 1 ns to the isomeric state ^{235}U^m at an excitation energy of 76.8 keV. It decays to the ground state ^{235}U^g by an E3 transition. As sketched to the right in Figure 21, to produce ^{235}U^m the inner surface of a sphere was covered with a thin layer of ^{239}Pu and following α-decay the recoiling isomers were collected on a target backing positioned at the centre of the sphere. Cross section measurements were performed for two neutron energies, ≈ 5 meV and ≈ 56 meV. At these energies the ratios of cross sections isomeric to ground state were found to be (1.61±0.44) and (2.47±0.45), respectively [56]. The isomeric state has hence a slightly larger fission cross section than the ground state.

9. Summary

In 35 years of fission research at the Institut Laue-Langevin a wealth of data on thermal neutron fission was accumulated and many new properties of this multi-faceted process were discovered. The merits of the outstanding performance of the High Flux Reactor and the ancillary equipment are obvious. Many if not most of the experiments reported in the foregoing could not have been performed elsewhere in the world.

Highlights of research were the study of super-asymmetric mass distributions, even-odd effects in the charge distributions, cold fission in the energy distributions, detailed investigations of ternary and quaternary fission, binary and ternary fission induced by polarized neutrons and, last not least, the measurement of fission cross sections.

There are evidently several properties of the fission process which were not tackled so far. There are the properties of prompt neutrons and gammas which deserve to be investigated at the limits of mass, charge and energy distributions. The insight having been gained at the limits of phase space should be complemented and corroborated by the characteristic of the deexcitation process. There is further the issue on angular momentum of fission fragments where much work is still to be done and could be done at the ILL.

References

1. H. Ewald et al., *Z. Naturforschg* **19a**, 194 (1963).
2. E. Moll et al., *Nucl. Instr. Meth.* **123**, 615 (1975).
3. G. Fioni et al., *Nucl. Instr. Meth.* **A 332**, 175 (1993).
4. P. Heeg et al., *Nucl. Instr. Meth.* **A 278**, 452 (1989).
5. J.P. Bocquet et al., *Nucl. Instr. Meth.* **A267**, 466 (1988).
6. A. Oed et al., *Nucl. Instr. Meth.* **205**, 451 (1983).
7. B.D. Wilkins et al., *Phys. Rev.* **C 14**, 1832 (1976).
8. I. Tsekhanovich et al., Proc. "Seminar on Fission", World Scientific 2004, p 56.
9. H.O. Denschlag et al., Proc. "Seminar on Fission", World Scientific 2000, p 5.
10. D. Rochman et al., *Nucl. Phys.* **A 735**, 3 (2004).
11. J.L. Sida et al., *Nucl. Phys.* **A 592**, 233c (1998).
12. I. Tsekhanovich et al., *Nucl. Phys.* **A 659**, 217 (1999).
13. J.P. Bocquet et al., *Z. Physik* **A 335**, 41 (1990).
14. W. Lang et al., *Nucl. Phys.* **A345**, 34 (1980).
15. G. Martinez et al., *Nucl. Phys.* **A 515**, 433 (1990).
16. P. Schillebeeckx et al., *Nucl. Phys.* **A 580**, 15 (1994).
17. D. Rochman et al., *Nucl. Phys.* **A 710**, 3 (2002).
18. M. Djebara et al., *Nucl. Phys.* **A 496**, 346 (1989).
19. H.-G. Clerc et al., *Nucl. Instr. Meth.* **124**, 607 (1975).
20. J.P. Bocquet et al., *Nucl. Phys.* **A502**, 213c (1989).
21. H. Nifenecker et al., *Z. Phys.* **A 308**, 39 (1982).
22. F. Reymund et al., *Nucl. Phys.* **A 678**, 215 (2000).
23. B.L. Tracy et al., *Phys. Rev.* **C 5**, 222 (1972).
24. J.L. Sida et al., *Nucl. Phys.* **A 592**, 233c (1998).
25. I. Tsekhanovich et al., *Nucl. Phys.* **A 688**, 633 (2001).
26. J. Kaufmann et al., Proc. "Nucl. Data for Science and Techn", Springer Berlin, 1992, p 131.
27. P. Schillebeeckx et al., *Nucl. Phys.* **A580**, 15 (1994).
28. J.C.D. Milton and J.S. Fraser, *Phys. Rev.* **111**, 877 (1958).
29. C. Signarbieux et al., *J.Physique Lettres* **42**, L437 (1981).
30. P .Armbruster et al., Proc. "Nuclei far from Stability", CERN 81-09, p 675.
31. P.Armbruster et al., *Rep. Prog. Phys.* **62**, 465 (1999).
32. C. Signarbieux, priv. communication 1990.
33. F. Gönnenwein and B. Börsig, *Nucl. Phys.* **A 530**, 27 (1991).

34. J. Trochon et al., Proc. "50 Years with Nucl. Fission", Am. Nucl. Society, 1989, p 313.
35. W. Schwab et al., *Nucl. Phys.* **A 577**, 674 (1994).
36. H.-G.Clerc, in "Heavy Elements and Related Phenomena", World Scientific 1999, p 451.
37. U. Quade et al., in "Lecture Notes in Physcis", Springer, 158, 40 (1982).
38. F.-J. Hambsch et al., *Nucl. Phys.* **A 726**, 248 (2003).
39. T.D.Thomas et al., Proc. "Physics and Chemistry of Fission", IAEA Vienna, 1965, p 467.
40. J.F. Berger et al., *Nucl. Phys.* **A 428**, 23c (1984) and Nucl. Phys. A 502, 85c (1989).
41. M. Hesse et al., Proc. "Dyn. Aspects of Nucl. Fission", JINR Dubna 1996, p 238.
42. I. Halpern, *Annual. Rev. Nucl. Science* **21**, 245 (1971).
43. M. Mutterer et al., Proc. "Fiss. and Prop. Neutron-rich Nuclei", World Scientific, 1998, p 119.
44. O. Serot et al., *Eur. Phys. J.* **A 8**, 187 (2000).
45. S. Vermote et al., *Nucl. Phys.* **A 837**, 176 (2010).
46. Yu. Kopatch et al., *Phys. Rev.* **C 65**, 044614 (2002).
47. U. Köster et al. Proc., "Seminar on Fission", World Scientific 2000, p 77.
48. C. Wagemans et al., *Nucl. Phys.* **742**, 291 (2004).
49. P. Jesinger et al., *Eur. Phys. J.* **A 24**, 379 (2005).
50. A. Kötzle et al., *Nucl. Instr. Meth.* **A 440**, 750 (2000).
51. P. Jesinger et al., *Nucl Instr. Meth.* **A 440**, 618 (2000).
52. A. Gagarski et al., AIP Conference Proceedings 1175, 2009, p 323.
53. F. Goennenwein et al., *Phys. Lett.* **B 652**, 13 (2007).
54. C. Wagemans et al., AIP Conf. Proc. 447, Woodbury, USA, 1998, p 262.
55. O. Serot et al., AIP Conf. Proc. 798, Melville, New York, 2005, p 182.
56. A. D`Eer et al., *Phys. Rev.* **C 38**, 1270 (1988).

MECHANISMS FOR LIGHT CHARGED-PARTICLE EMISSION DURING NUCLEAR FISSION

NICOLAE CARJAN

Horia Hulubei National Institute for Physics and Nuclear Engineering
P.O.Box MG-6, Bucharest, Romania
and
CENBG, CNRS/IN2P3-University Bordeaux I,
33175 Gradignan Cedex, France

A retrospective of some theoretical advances made in this field during the last 40 years is given. The following mechanisms are discussed: the adiabatic release, the sudden release, the evaporation from primary fragments, the α-decay during the descent from saddle to scission, the double neck-rupture and the uncertainty principle. Analyses of N_α/N_F ratios for various nuclei confirms the important role played by the spectroscopic factor, i.e. by the probability that an α-cluster is preformed in a fissioning system. The observed strongly peaked emission perpendicular to the fission axes, that is believed to be an indication of the release from the neck region, can be obtained with a continuous distribution of the emission points along the entire nuclear surface. This is true both in the case of a classical emission over the Coulomb barrier and in the case of an emission through quantum tunneling.

1. Introduction

Investigations in the field of nuclear binary fission have elucidated various aspects concerning the transition of the nucleus from the ground state to the last saddle point, especially through studies of shape isomers and of narrow intermediate structure in induced fission cross sections near threshold. On the other hand, comparatively little is known about the descent of the nucleus towards the scission point, and the character of this descent constitutes at present a highly debated aspect of nuclear fission. From this fact it is evident that any related phenomenon furnishing information about the last fission stage is particularly interesting.

The light-particle-accompanied fission (Fig. 1) represents apparently an excellent example of such phenomenon since the light-particle release occurs both in space and in time near the scission point. In 95% of cases the light particle is an α-particle. How has one tried to extract the information, about the manner in which the nucleus breaks, which these particles carry along? One usually assumed simple distributions (gaussian) for the quantities defining the

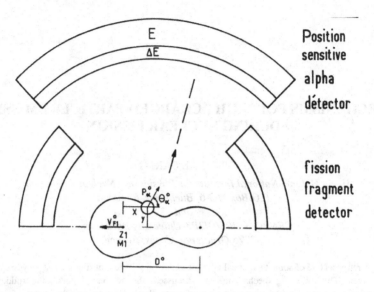

Figure 1. Experimental configuration and initial parameters.

nuclear configuration at the moment of the α-particle emission (see Fig. 1 for their meanings) and scaled them to the experimental distributions via the calculation of the trajectories of the α particle and of the two fragments in their mutual (mainly Coulomb) field. These initial distributions were considered to be independent of each other and the values of the initial parameters were therefore sampled randomly inside their ranges. The number of initial parameters can be slightly reduced from general considerations like energy conservation and uncertainty principle. The experimental mass distribution corrected for neutron emission can also be used to fix W(M1) and eventually W(Z1) by assuming N/Z = const during fission. However the available experimental results are neither totally independent nor have enough structure to allow the unique determination of the remaining distributions. The first to have pointed out that it is impossible to obtain unique values of the initial parameters was A. Gavron [1] for the spontaneous fission of ^{252}Cf.

In conclusion, to remove this ambiguities one should try to:
a) increase the information contained in the experimental data by revealing possible structures through improved resolution and statistics and
b) provide theoretical estimates for some of the initial parameters by understanding the emission mechanism of the scission α-particles.

Some progress made in the latter direction during the last 40 years is presented in the next sections.

2. The adiabatic release

The most naïve picture of the light particle release is a slow thinning of the necks on both sides of the particle. However the simultaneous vanishing of the two necks was considered improbable and this model was abandoned. It was argued that, if the vanishing is not simultaneous, the nuclear force from the other side will absorb the particle.

More recent dynamical calculations in the frame of the finite-size liquid-drop model have shown [2] that, when sufficiently heavy nuclei (A> 300) undergo fission, they do not lead to compact shapes but they are forming a rather long cylindrical neck which subsequently contracts at the extremities to divide the nucleus into three fragments (Fig. 2). This behaviour can be

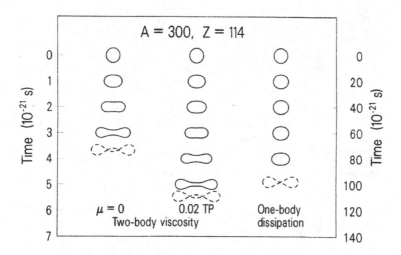

Figure 2. Time evolution of nuclear shapes leading to true ternary fission.

understood in terms of the well known hydrodynamical instability [3,4] of a fluid cylinder of radius a against a perturbation with $\lambda > 2\pi a$ (Plateau - Rayleigh). Lighter nuclei, for which the maximum of the fission probability evolves in the binary valley, have a certain chance (given by the tail of the probability distribution that extends to double-neck shapes) to fission into three fragments.

For low-energy fission the shell effects will modify the masses of the three fragments favoring the α particle as middle fragment. Cluster formation on the top of long-neck instability represents therefore a possible origin for the fission α particles.

3. The sudden release

The idea of the sudden release was proposed by Halpern [5]. Due to the very fast disappearance of the neck connecting the fission fragments ' just before scission', the potential of the interaction between an α- cluster and the fissioning nucleus changes so rapidly that the cluster has no time to follow the change. In this way, 'immediately after scission', a part of α- wave function (namely that located between the fragments) will find itself outside the range of the nuclear attraction and therefore released. The sudden approximation is only justified if the typical neck-rupture time is very small compared to the typical neck-crossing time of an α- particle.

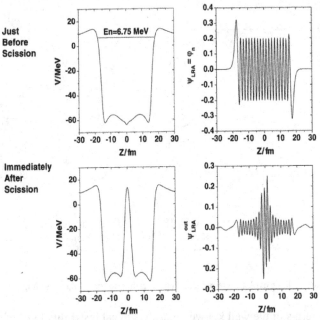

Figure 3. The potential change for an α particle due to the sudden transition (left column) together with the total α wave function (up-right) that is unchanged during the transition and its unbound components (down-right).

In Ref [6] the sudden release was developed quantitatively along the deformation axis z. To illustrate this calculation we present, in the left column of Fig. 3, the sudden change in the potential and in the right column the total α- wave function and its unbound part. Although one-dimensional, this model reproduced the N_α/N_f values, accounted qualitatively for the angular distribution (including the equatorial and the polar components) and showed the strong dependence of the energy transfer probability on the nuclear elongation at scission in agreement with the experimental evidence [7].

4. The alpha-decay between saddle and scission

This hypothesis started from the observation that all fissile nuclei are α emitters in their ground state. This property (of being an α emitter) does not cease when the nucleus undergoes fission but it is, of course, continuously influenced by the new shapes and the new energy balances that the nucleus goes through on its way to scission. A possible source for the scission α's would therefore be the α-decay from the last stage of the fission process [8,9].

The simplest mechanism, through which an α cluster in a fissioning nucleus can gain enough energy to escape, is the collision with the moving wall of the α-nucleus potential. An example of barriers through which an α particle in ^{236}U must tunnel, at each hit, as it moves along the ρ-axes is shown in Fig. 4.

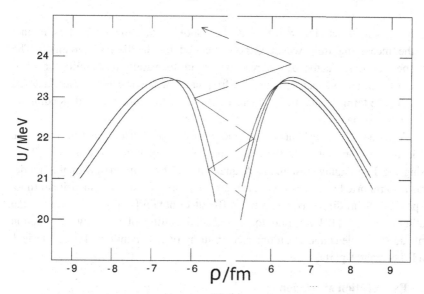

Figure 4. Barriers seen by the α-particle in ^{236}U and α-energies during the last reflexions.

This schematic picture leads to the following basic equations:
1) The energy spectrum at the moment of emission:

$$N(Q) = \sum_i W_i T^i(Q) \prod_{n=1}^{n_Q^i} [1 - T^i(E_n)]$$

where n_Q is the number of times the particle attempts to escape before it

reaches the energy Q, T(Q) is the tunneling probability and the index i represents a given fission mode. The distribution calculated according to this equation is slightly asymmetric in agreement with the experimental data [10].
2) The rate of emission:

$$N_\alpha/N_f \neq \lambda_\alpha/\lambda_f$$

since we are not in a branching situation (no dual decay). If fission and α decay occur simultaneously, as in this model,

$$N_\alpha/N_F = \int_0^\tau \lambda_\alpha(t)\omega_t dt \quad for \ \lambda_\alpha\tau \ll 1$$

$$= 1/\hbar * \delta_\alpha^2 * \int_0^\tau T[Q(t)] \prod_{n=1}^{n_i} [1 - T(E_n)] \, dt$$

for a single scission mode. τ is the saddle-to-scission time; it is in fact the measuring time which is now dictated by the fission dynamics. The spectroscopic factor δ_α^2 represents the preformation probability of an α cluster in the nuclear surface. The frequency of this process is thus factored into one term depending on nuclear structure and another term depending on nuclear dynamics.

A 'one-body' mechanism for the transfer of kinetic energy from the collective fission-motion to the cluster motion was quantitatively studied [9,11] using the Lagrangian formalism. A quantum mechanical version of this model was also reported [12]. However accurate solutions of the two-dimensional time-dependent Schrodinger equation are difficult to obtain [13]. For this reason the authors of Ref. [12] resorted to an artificial cutting of the α wave function outside the nucleus and to an untested splitting of the Hamiltonian into 'vertical' and 'horizontal' parts.

5. Evaporation at scission

The characteristics of the fission fragments and of the ternary particle are taken from the statistical theory of nuclear fission. Trajectory calculations of three point-charges are then performed using these predictions as initial conditions and the over-all energy and angular distributions are reproduced [14,15]. As mentioned in the introduction, this procedure being highly ambiguous, doesn't ensure the correctness of the evaporation hypothesis.

It is also impossible to reconcile this model with certain observation such as:

- for the same compound nucleus, there are more α's in spontaneous fission than in neutron-induced fission.
- for the same (n,f) reaction, the ratio N_α/N_F is slightly decreasing with increasing incident energy.

Moreover, a statistical theory would predict more scission protons than α's and a mainly polar emission (since the Coulomb barrier is lower in these cases).

6. Double random neck-rupture

This model improves the naïve picture (Sec. 2) of simultaneous rupture of the neck in two places by discussing the time scale at which two non-simultaneous ruptures can occur. It makes the hypothesis that during the lifetime of the neck 't_n', two independent ruptures occur with equal probabilities. These probabilities are uniformly distributed in space and time: they are the same at each point along the neck and at each moment of time during 't_n' [16,17]. In addition to t_n there are two other times (see Fig. 5) that are important: t_r (of the neck rupture) and t_{abs} (of the neck absorption by the fragments). Consequently, if the 2nd rupture arrives in the interval $\{t_r + t_{abs} - t_r\}$ after the 1st rupture the fission is ternary. The ratio ternary over binary fission should be proportional with the ratio of the times involved: $T/B = const\ (t_{abs} / t_r)$.

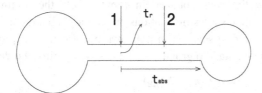

Figure 5. Schematic representation of the neck-rupture and neck-absorption times.

At present, our lack of knowledge of the complicated dynamics of the neck rupture does not allow a reliable estimate of these times.

7. The uncertainty principle

There are different ways to formulate the quantum mechanical uncertainties:

1) $\Delta\rho\Delta p_\alpha = \hbar/2$

$\Delta E_\alpha = (p_\alpha/m_\alpha)\ \Delta p_\alpha = (2E_\alpha/m_\alpha)^{1/2}\ \hbar/(2\Delta\rho) = 2.3(E_\alpha)^{1/2}/\Delta\rho$

2) $\Delta E\ \Delta t = \hbar/2$

$\Delta E_\alpha = 3.3/\Delta t$

$[E] = MeV,\ [\rho] = fm,\ [t] = 10^{-22}\ sec$

If the particle is squeezed in the neck region between the walls of the potential of its interaction with the rest of the nucleus, it is confined to a very small $\Delta\rho$ and has a very large energy distribution. According to the formulation 1) its tail can approach the top of the barrier. Likewise, if the particle has to be emitted in a very short interval of time it will have again a very broad energy distribution according to 2). Although it is difficult to assert the effect of the uncertainty principle precisely, one thing is sure: if it has any influence on the emission we cannot get around it.

8. Important role played by the clustering probability

In Sec. 4 we expressed the α multiplicity (the number of α particles emitted per fission event) as a product between the α spectroscopic factor and the integrated time-dependent probability that an α particle escapes at the top of the barrier through quantum tunneling or over-transmission. The time dependence of this escape probability is a consequence of the coupling between the α particle and the fission degrees of freedom. It is therefore dictated by the fission dynamics between saddle and scission and it should scale with the fissility parameter as the liberated distortion energy does. This was elegantly demonstrated in Refs. [7,18] where $(N_\alpha/N_F)/\delta_\alpha^2$ was plotted versus Z^2/A of the fissioning nucleus and a smooth increasing function was obtained in contrast with the original curve, (N_α/N_F) versus Z^2/A, which shows fluctuations due to the shell structure.

The value of δ_α^2 close to scission in unknown but it can be approximately estimated from the ground-state to ground-state α decay experimental data:

$$\delta_\alpha^2 = b \times (\lambda_{exp}/\lambda_{WKB})$$

where b is the branching ratio. The success of the above mentioned analyses also indicates that the relative variation of δ_α^2 with nuclear deformation is approximately the same for all nuclei involved.

9. Emission always perpendicular to the deformation axis

The observed main direction of emission of the fission α particles almost perpendicular to the fission axis was considered an evidence for their emission from the neck region. In this section we will demonstrate that this inference is false and that even the opposite assumption (constant emission probability from all the points of the nuclear surface) leads to an angular distribution strongly peaked at 84° with respect to the light fragment.

Finite-size trajectory calculations [19] predict a deflection function $\theta_{\alpha L}(z_\alpha^0)$ of oscillating shape as exemplified in Fig. 6. The trajectories are therefore focused at angles corresponding to the extrema of this function: namely 81° and

89°. Using the analogy with the classical collision theory:

$$\frac{d\sigma}{d\theta} = b\left(\frac{d\theta}{db}\right)^{-1}$$

where θ (b) is the deflection function and supposing that the α particles are emitted with constant probability from any point along the barrier ridge, one can calculate the differential cross section as:

$$\frac{d\sigma}{d\theta_{\alpha L}} = \sum_k \rho(z_k)\left(\frac{d\theta_{\alpha L}}{dz}\right)^{-1}\bigg|_{z=z_k}$$

The result is represented with a dotted curve in Fig. 6.

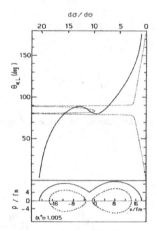

Figure 6. Final angle $\theta_{\alpha L}$ between the α particle and the light fragment direction (top) for different initial α positions all around ^{236}U at scission (bottom).

To compare with the experimental data one has to fold with the angular resolution function:

$$\frac{d\sigma}{d\theta}\bigg|_{\theta=\theta_0} = \int_{-\infty}^{\infty} \frac{d\sigma}{d\theta} r(\theta,\theta_0)\, d\theta$$

where

$$r(\theta,\theta_0) = \frac{1}{\sqrt{2\pi}\epsilon} \exp\left[-\frac{(\theta-\theta_0)^2}{2\epsilon^2}\right]$$

For an angular resolution $\epsilon < 5°$, the resulting angular distribution presents a structure of the type shown in Fig. 7. An experiment looking for such structures would also require a good mass and energy resolution to allow a certain selection of scission configurations. To avoid the blurring of the angle $\theta_{\alpha L}$ by neutron emission one should also select cold events.

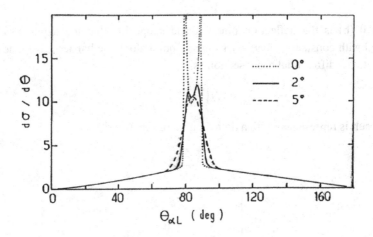

Figure 7. Calculated α particle angular distribution corresponding to the deflection function from Fig. 6 for different angular resolutions.

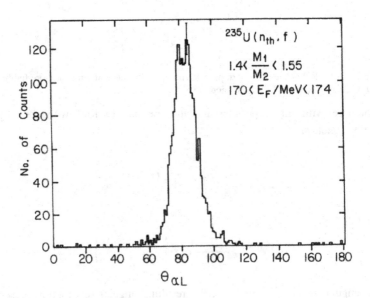

Figure 8. Angular distribution for α particles measured [20] in coincidence with fission fragments having well defined kinetic energy and mass ratio.

Measured angular distributions for events selected according to the above mentioned criteria seem to present such structures as can be seen in Fig. 8. Unfortunately there is not enough statistics to confirm or disprove this still attractive possibility.

Finally we should point out that emission of s-state protons by quantum tunneling from very deformed nuclei always leads to an angular distribution perpendicular to the deformation axis [21].

10. Conclusions

Light charged-particle emission during nuclear fission was and still is a very fertile field to develop our imagination. Many bright ideas have been advanced but all came against our lack of detailed knowledge of the dynamics of large amplitude collective motions, and in particular of the diabatic dissipative dynamics of the last stage of nuclear fission.

Acknowledgments

Work partially supported by the project PN09370102
I am grateful to Yannick for his help with Figs. 4 and 5.

References

1. A. Gavron, *Phys. Rev.* **C11**, 580 (1975).
2. N. Carjan, A. J. Sierk and J. R. Nix, *Nucl. Phys.* **A452**, 381 (1986).
3. S. Chandrasekhar, *Hydrodynamic and Hydromagnetic Stability* (Claredon, Oxford, 1961) pp.515-516, 537-544.
4. V. Brossa and S. Grossmann, *Worskhop on Gross Properties of Nuclei and Nuclear Excitations XI*, Hirscheg,, ISSN 0720-8715 (1983) p. 187.
5. I. Halpern, *First Symposium of Physics and Chemistry of Fission*, Salzburg, 1965 (IAEA, Vienna, 1965) vol II, p.369.
6. O. Serot, N. Carjan and C. Wagemans, *Eur. Phys. J.* **A8**, 187 (2000)
7. O. Serot and C. Wagemans, *Nucl. Phys.* **A641**, 34 (1998).
8. N. Carjan, *J. Physique* **37**, 1279 (1976).
9. N. Carjan, *Ph.D. Thesis,* TH Darmstadt, Difo-Druck, Bamberg (1977).
10. M. Mutterer, Yu. N. Kopatch, et al., *Phys. Rev.* **C78**, 064616 (2008).
11. S. Oberstedt and N. Carjan, *Z. Physik* **A344**, 59 (1992).
12. O. Tanimura and T. Fliessbach, *Z. Physik* **A328**, 475 (1987).
13. M. Rizea and N. Carjan, *Computer Physics Research Trends*, Nova Science Publishers, New York (2007), pp. 185-210.
14. P. Fong, *Phys. Rev.* **C3**, 2025 (1971).
15. J. Lestone. *IJMP* **E17**, 323 (2008).

16. V. Rubchenya, *Sov. J. Nucl. Phys.* **35**, 334 (1982).
17. V. Rubchenya and S. Yavshits, *Z. Physik* **A329**, 217 (1988).
18. S. Vermote et al., these Proceedings.
19. N. Carjan and B. Leroux, *Phys. Rev.* **C22**, 2008 (1980).
20. J. P. Theobald, M. Mutterer, J. Panicke et al., private communication.
21. P. Talou, N. Carjan and D. Strottman, *Nucl. Phys.* **A647**, 21 (1999).

Characteristics of the Fission Process

NEW DEVELOPMENTS IN THE MICROSCOPIC DESCRIPTION OF THE FISSION PROCESS

H. GOUTTE[†]

Grand Accélérateur National d'Ions Lourds (GANIL), CEA/DSM-CNRS/IN2P3, BP 55027, F-14076 Caen Cedex 5, France

R. BERNARD

CEA, DAM, DIF, F-91297 Arpajon, France

Fission is a complex process which highlights many nuclear properties. Among the different theoretical approaches able to describe fission, microscopic ones have the advantage of describing the nuclear structure and the dynamics in a consistent manner. Along this line, we are now developing a formalism able to treat on the same footing the collective dynamics and the intrinsic excitations. This approach is based on the non adiabatic time-dependent Generator Coordinate Method, where couplings between HFB states and 2 quasi-particle states are explicitly taken into account. Guidelines of the new formalism under development are presented and some preliminary results on overlaps between non excited and excited states are discussed.

1. Introduction

From a theoretical point of view, the description of the fission process is at the crossroads of many research topics of nuclear physics, fission bringing into play both structure properties of nuclei and dynamical properties. Fission appears as a large amplitude collective motion, in which a large rearrangement of the internal structure of the nucleus is observed during the process with a crucial role played by the shell effects and the correlations.

A major challenge in theoretical nuclear physics is the development of a consistent approach able to describe on the same footing the whole fission process, i.e. properties of the fissioning system, fission dynamics and fission fragment distributions. As a first step, a microscopic time-dependent and quantum mechanical formalism has been developed based on the Gaussian Overlap Approximation of the Generator Coordinate Method with the adiabatic

---

[†] goutte@ganil.fr

approximation [1]. Results obtained for the low-energy fission of ^{238}U [2] encouraged us to pursue further studies of fission along these lines with some additional improvements. For instance, at higher energies, a few MeV above the barrier, the adiabatic approximation doesn't seem valid anymore. Taking the intrinsic excitations into account during the fission process will enable us to determine the coupling between collective and intrinsic degrees of freedom, in particular from saddle to scission. Here the extended formalism under development is first sketched. Preliminary results on overlaps between deformed excited and non excited basis states are then discussed.

2. Hierarchy of the correlations in the mean-field based approaches

The main advantage of "mean-field based approaches" comes from the hierarchy of the correlations that are taken into account, with first short range correlations, then pairing, multipolar vibrations, particle-vibration coupling ... The most commonly used approaches are displayed in Table 1. First, at the mean field level the initial many body problem is simplified, considering that each particle interacts with a mean field generated by all the particles. Thus, Hartree Fock solutions (HF) are particle or hole states, i.e. states completely filled or empty. In open shell nuclei, in which the last occupied level is not completely filled, the part of the residual interaction that first enters into play is the pairing interaction. This extension is achieved with BCS or Hartree-Fock-Bogoliubov (HFB) approaches, and the independent quantities are the quasi-particles.

In beyond mean-field approaches, oscillations of the field due to a residual interaction are taken into account. There are two types of approaches which incorporate the long range correlations missing at the mean-field level: (Quasi-Particle) Random Phase Approximation (Q)RPA-type approaches and approaches such as the Generator Coordinate Method (GCM). On the one hand, (Q)RPA methods deal with small amplitude correlations and introduce the coupling between HF(B) ground state and particle hole excitations (or more generally 2 quasi-particle excitations for the QRPA). They give access to the correlated ground state and to low and high energy modes. They are particularly well adapted for the study of magic and semi-magic nuclei, in which the small amplitude approximation is well justified. On the other hand, the Generator Coordinate Method incorporates large amplitude vibrations by considering a continuous mesh on collective variables and by mixing the configurations along these variables. This method can be generalized in principle to N degrees of freedom, but in practice calculations are hardly feasible when more than two are treated simultaneously.

In order to take into account dynamical properties of the fission process, the time-dependent Generator Coordinate Method (TDGCM) appears to be very powerful [1,2]. In this approach, i) collective and intrinsic degrees of freedom are assumed to have different time scales, ii) the adiabatic hypothesis is used and iii) the dynamical evolution of the system is solved in a restricted space generated by collective variables.

Table 1. Approaches based on the mean-field and hierarchy of the correlations (* proton-neutron pairing is not considered here)

Correlations	Short range	Long range		
			"Collective" correlations	
Name	Short range correlations	Usual pairing*	Small amplitude	Large amplitude
Origin	Short range repulsive part of the nuclear force	Coupling between HF state and 2p2h S=0 T=1	Coupling between HFB state and 1p1h states	Shape coexistence
Wave function (w.f)	Independent particles	Independent quasi-particles	w. f more general than a w. f of independent particles or quasi-particles	
Method	Hartree-Fock (HF) theory with an effective force	Hartree-Fock-Bogoliubov (HFB)	RPA/QRPA	GCM

However, it is well known that collective modes and intrinsic excitations should be coupled. We have then developed a new formalism based on the GCM+quasi-particle excitations to explicitly treat non-adiabatic motion i.e. collective modes, intrinsic excitations and their couplings.

Let us note that the D1S force [3,4] is well adapted for such a formalism. Indeed, thanks to the finite ranges in the central term, pairing and mean-field correlations are treated on the same footing, and in addition the D1S force has revealed to be able to take into account excitations beyond the mean-field [5,6].

3. Formalism

In the non adiabatic time-dependent Generator Coordinate method, the nuclear state is defined as:

$$|\Psi(t)\rangle = \sum_i \int dq\, f_i(q,t)|\Phi_i(q)\rangle, \qquad (1)$$

where the time-dependent weight function $f_i(q,t)$ are obtained through

$$\frac{\partial}{\partial f_i^*(q,t)} \int_{t_1}^{t_2} \langle \Psi(t)|\hat{H} - i\hbar\frac{\partial}{\partial t}|\Psi(t)\rangle dt = 0. \qquad (2)$$

In Eq. (1), $\hat{H}$ is the nuclear Hamiltonian, and $|\Phi_i(q)\rangle$ is a set of static states depending on the collective coordinates q and on single particle excitations i, where i = 0 stands for the lowest-energy state.

The variational principle (Eq. (2)) leads to the generalized Hill and Wheeler equation:

$$\sum_i \int dq (\langle \Phi_j(q')|\hat{H}|\Phi_i(q)\rangle - i\hbar \langle \Phi_j(q')|\Phi_i(q)\rangle) f_i(q,t) = 0. \qquad (3)$$

In the usual adiabatic case, without intrinsic excitations, the adiabatic nuclear state $|\Psi_{adiab}(t)\rangle$ is defined as,

$$|\Psi_{adiab}(t)\rangle = \int dq\, f_{adiab}(q,t)|\Phi(q)\rangle, \qquad (4)$$

where $|\Phi(q)\rangle$ is the lowest energy state at a fixed q obtained from the constrained Hartree-Fock-Bogoliubov method. In [1,2], the GCM problem is solved using the Gaussian Overlap Approximation (GOA) [7], and the Hill-Wheeler equation is reduced to a Schrödinger equation:

$$\hat{H}_{coll} g(q,t) = i\hbar \frac{\partial g(q,t)}{\partial t}, \qquad (5)$$

with g(q,t) the gauss transform of $f_{adiab}(q,t)$ and

$$H_{coll} = -\frac{\hbar^2}{2} \sum_{i,j} \frac{\partial}{\partial q_i} B_{ij}(q) \frac{\partial}{\partial q_j} + V(q) + \sum_{i,j} ZPE_{ij}(q). \qquad (6)$$

In Eq. (6), B(q) is the inverse of the collective mass and ZPE is the zero point energy correction. The Gaussian Overlap Approximation appears as a very practical tool to simplify the initial very complicated integro-differential Hill and Wheeler equation into a Schrödinger equation.

Here the problem deals with the generalization of the GOA when intrinsic excitations are taken into account in the formalism. To solve this problem, the excitations along the fission paths have been first studied and the overlaps between excited states have then been calculated and compared with the overlaps between HFB lowest-energy states.

4. Generalized overlap: preliminary results

Overlaps $\langle \Phi(q) | \Phi(q') \rangle$ between HFB states at different deformations are plotted in the upper panel of Fig. 1 as functions of q-q' for different (q+q')/2 values. A detailed study of their shapes is in progress. We note that the width of the overlap depends on the deformation. The smallest width is found for (q+q')/2 = 5000 fm^2 and the largest one for 14000 fm^2.

For the excited states, K=0 two-quasi-particle (2qp) excitations coupled to K=0 have been selected. Calculations have been performed in ^{236}U. First results show that the lowest neutron or proton 2qp excitations are lying in the range 1MeV- 3 MeV. Thus, these excitations may be low enough to play a role during the fission process. Indeed, these results have to be compared with those obtained in the experiment of ^{238}U and ^{239}U where proton pair breakings are observed for excitation energy of 2.3 MeV above the barrier [8–10].

Non diagonal overlaps $\langle \Phi(q) | \Phi_i(q') \rangle$ with i corresponding to different K=0 excitations built on two proton quasi-particle states with K=1/2$^-$, 3/2$^-$ and 5/2$^-$ are plotted on the lower panel of Fig.1. The maximum of the overlaps is predicted to be very small, namely 0.08 compared to 1 for the diagonal terms (upper panel).

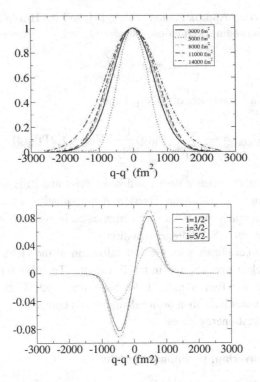

Fig 1. Upper panel: overlaps $\langle \Phi(q)|\Phi(q')\rangle$ calculated at different (q+q')/2 deformations in ^{236}U. Lower panel: overlaps between HFB minima and K=0 excited states based on two proton quasi-particle states with K = $1/2^-$, $3/2^-$, and $5/2^-$ at (q + q')/2 = 5000 fm^2.

A more precise study of all these overlaps is in progress. Such a study is very important to have a clear protocol to build a Schrödinger equation from the Hill and Wheeler equation. Results will be discussed in detail in [11].

5. Conclusion

A study of the role played by the intrinsic excitations along the fission paths has been undertaken. Preliminary results on norm overlaps between non-excited and excited 2 qp states have been presented in ^{236}U. The next step is now to calculate the energy kernel between different basis states. On the long range, we expect to develop a fully microscopic non adiabatic formalism to improve results on fission fragment properties such as fragment charge and mass distributions.

References

1. 1.J.-F. Berger, M. Girod and D. Gogny, *Nucl. Phys.* **A428**, 23c (1984).
2. 2.H. Goutte, J.-F. Berger, P. Casoli and D. Gogny, Phys. Rev. C71, 024316 (2005).
3. 3.J. Dechargé and D. Gogny, *Phys. Rev.* **C21**, 1568 (1980).
4. 4.J.-F. Berger, M. Girod and D. Gogny, *Comp. Phys. Comm.* **63**, 365 (1991).
5. S. Péru and H. Goutte, *Phys. Rev.* **C77**, 044313 (2008).
6. J.-P. Delaroche, M. Girod, J. Libert, H. Goutte, S. Hilaire, S. Péru, N. Pillet and G.-F. Bertsch, *Phys. Rev.* **C81**, 014303 (2010).
7. P. Ring and P. Schuck, *The Nuclear Many Body Problem*, (Springer-Verlag, New York, 1980).
8. S. Pommé et al, *Nucl. Phys.* **A572**, 237 (1994).
9. F. Vives, F.-J. Hambsch, H. Bax and S. Oberstedt, *Nucl. Phys.* **A662**, 63 (2000).
10. S. Pommé et al., *Nucl. Phys.* **A560**, 689 (1993).
11. R. Bernard et al., to be submitted.

QUANTUM ASPECTS OF LOW-ENERGY NUCLEAR FISSION

W. FURMAN

Joint institute for nuclear research, Dubna Moscow region, 141980, Russia,
furman@nf.jinr.ru SECOND

A helicity representation for fission product channels with correctly defined parity is used to describe neutron induced fission with arbitrary spin density matrix in ingoing channel. Recently obtained data for ROT effect in binary fission give evidence for high accuracy of the helicity representation just at scission. A general expression for differential cross-section of (n,f)-reaction is obtained. In the framework of multilevel, many channel R-matrix theory the reduced S-matrix for JΠK effective channels rigorously derived. These channels include fission modes in natural way. Theoretical analysis of experimentally observed P-even and P-odd interference effects in low energy nuclear fission allows one to make some essential conclusions on basic mechanism of the process.

1. Introduction

Resonance neutron induced fission or spontaneous fission goes through or from the states of fissioning system with fixed parity Π (excluding rather small parity violating P-odd effects), total spin J and its projections K onto deformation axis. High energy resolution and high intensity of pulsed neutron beams allow one to study in experiment rather small and delicate effects related with neutron resonances.

In particular it has been observed an inter-resonance interference in total and differential fission cross sections, as well as small variations (over resonances) of TKE&mass distributions, prompt neutron yields and even space parity violation. An use of polarized neutrons and aligned or polarized target nuclei allows one to observe various P-even and P-odd angular correlations of fission fragments (FF).

All these experimental observations reveal a quantum-mechanical nature of fission process. As it will be shown below this means that nuclear fission is governed basically by the wave function (WF) of transition state (TS) which describes the possible "trajectories of movement" in "deformation" (configuration) space. The transition state WF is a" carrier" of nuclear shell effects which define the "menu" of accessible fission mode at scission as well as essential characteristics of respective fission barriers. The symmetry properties of TSWF ensure a reduction of FF channels to very limited numbers of "effective fission channels" attributed by quantum numbers JΠK. The following consideration will be based on some key assumptions.

The first is the adiabatic character of fission process. It expects that internal nucleon motion in fissioning nucleus is much "faster" than the collective one (the change of nuclear shape). So it is possible to use for TS WF the collective model wave function [1]. And its internal WF could be defined at any points of the way in deformation space.

The second is a conservation of the projection K on the "way in deformation space" starting from the compound state formed after neutron capture up to scission point. In other words it means that fissioning nucleus keeps its axial symmetry up to splitting for two (or more) FF. As it will be shown below the existence of P-even and P-odd angular correlations of FFs implies that the axial symmetry and adiabatic character of fission process preserves with good accuracy up to scission.

The third assumption is related to the role of shell effects in nuclear fission. There are well known arguments [2] following from the liquid drop approach why scission took place at very large deformation of fissioning nucleus. But from microscopical point of view the shell structure of heavy nuclei implies that they get possibility to split for FF only when the preformation factors (PF) of future FF in fissioning nucleus achieve reasonably large values [3].

As well just shell structure of fissionning nucleus is responsible for bifurcations of most probable trajectories in deformation space and defines main characteristics of fission modes and respective fission barriers.

Further consideration is based on the theoretical approach to description of nuclear fission developed in [4-7] and on use of semi-phenomenological arguments following from experimental observations. The consideration utilizes a formal theory of nuclear reactions and does not involve any dynamical models. Below most of theoretical results is presented without detailed derivation just in its final shape. Details could be found in the published papers [4-7].

2. Elements of formalism

The main spin and kinematics characteristics of fissioning nucleus and FF channel are shown in Fig. 1. A projection M of total spin J onto z-axis of laboratory frame is preserved during all fission process. The K is defined in internal coordinate system of fissioning nucleus. Spins of FF compose the channel spin $F = J_1 + J_2$ that in turn together with orbital momentum L forms the total spin $J = F + L$. Projections K_1, K_2 and κ of spins J_1, J_2 and F onto the fission axis r_f are the helicity quantum numbers. It is evident that the projection of spin J onto vector r_f is $\kappa = K_1 + K_2$ because $(L\, r_f) = 0$. Such very useful representation for FF channel had been introduced for the first time in ref. [8].

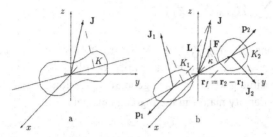

Fig.1. Spin and kinematics characteristics of binary nuclear fission.

For the general case of only axial symmetry of fissioning nucleus TS WF is expressed [1,6] ($K > 0$)

$$\Psi_f^{J\Pi KM}(\omega,\beta,\tau) = i^{\frac{1-\Pi}{2}} \sqrt{\frac{2J+1}{8\pi}} (D_{MK}^J(\omega)\Phi_K(\beta_f,\tau) + \Pi(-1)^{J+K} D_{M-K}^J(\omega)\Phi_{-K}(\beta_f,\tau)) \quad (1)$$

The angles ω describe the orientation of deformation axis in the laboratory reference system and the variables β_f and τ are related to the shape of fissioning nucleus and its internal motion correspondingly. The internal WF $\Phi_K(\beta_f,\tau)$ could be presented [9] as an expansion over fission modes m

$$\Phi_K(\beta_f,\tau) = \sum_m \alpha_m^K(\beta_f) \Phi_m^K(\beta_f,\tau) \quad (2)$$

Here the functions $\alpha_m^K(\beta_f)$ describe the relative contributions of fission modes m along the trajectory of fissioning system in multi-dimensional deformation space including the bifurcations caused by appearance of new type of symmetry of fissioning nucleus.

The wave function of FF channel c_f in helicity representation with definite parity Π could be written in the form [6]:

$$\varphi_{F|\kappa|\Pi JM}^{c_f}(\mathbf{n}_f) = i^{(1-\Pi\pi_1\pi_2)/2}(\varphi_{F|\kappa|JM}^{c_f}(\mathbf{n}_f) + \Pi\pi_1\pi_2(-1)^{J-F}\varphi_{F-|\kappa|JM}^{c_f}(\mathbf{n}_f)) = $$
$$= i^{(1-\Pi\pi_1\pi_2)/2}\sqrt{(2J+1)/8\pi}(D_{M|\kappa|}^J(\mathbf{n}_f)\chi_{F|\kappa|}^{c_f} + \Pi\pi_1\pi_2(-1)^{J-F} D_{M-|\kappa|}^J(\mathbf{n}_f)\chi_{F-|\kappa|}^{c_f}) \quad (3)$$

where

$$\varphi_{F\kappa JM}^{c_f}(\mathbf{n}_f) = \sum_L (-i)^L \sqrt{\frac{2L+1}{2J+1}} C_{F\kappa L0}^{JK} \varphi_{LFJM}^{c_f}(\mathbf{n}_f)$$

$$\varphi_{LFJM}^{c_f}(n_f) = \sum_{nM_1M_2} C_{FvLn}^{JM} C_{J_1M_1J_2M_2}^{Fv} \chi_{J_1M_1}^{c_f} \chi_{J_2M_2}^{c_f} i^L Y_{Ln}(n_f)$$

$$\chi_{F\kappa}^{c_f} = \sum_{M_1M_2} C_{J_1M_1J_2M_2}^{F\kappa} \chi_{J_1M_1}^{c_f} \chi_{J_2M_2}^{c_f}$$

The wave functions of FF $\chi_{J_iM_i}^{c_\alpha}$ are similar to TS WF only with good R-symmetry [1]. Symbols π_i indicate the parity of FF state. Using the above definitions it is possible to obtain a most general expression for (n,f)-reaction [6] with arbitrary spin density matrix in the ingoing channel:

$$\frac{d\sigma^{c_f}}{d\Omega_f} = \pi\lambda^2 \sum_{J'J} (g_{J'}g_J)^{1/2} \sum_{l'j'lj} \sum_Q \Phi_{l'j'ljJ'J}^Q (n_f n_k n_l n_s) B_Q^{c_f}(l'j'lj;J'J), \qquad (4)$$

where the kinematical factor Φ^Q (an explicit expression is presented in ref. [5]) depends on the relative orientation of four unit vectors n_f, n_k, n_l and n_s. Here n_k is directed along the collision axis when n_l and n_s are indicated the directions of target nucleus and incident neutron spins, respectively. Explicit form of the factor $B_Q^{c_f}$ is following

$$B_Q^{c_f}(l'j'lj;J'J) = \sum_{\Pi\Pi'} (1+\Pi'\Pi(-1)^Q)/2 \times (\sum_F i^{J-J'} C_{J0Q0}^{J'0} S_{J'}^*(l'j' \to F0\Pi'c_f) S_J(lj \to F0\Pi c_f) +$$
$$+ i^{(\Pi'-\Pi)\pi_1\pi_2/2} \sum_{F\kappa>0} C_{J\kappa Q0}^{J'\kappa} S_{J'}^*(l'j' \to F\kappa\Pi'c_f) S_J(lj \to F\kappa\Pi c_f))$$

Here S-matrix S_J includes all dynamics of the process. But the above differential cross section is practically unobservable. In experiment a summation over many FF channels c_f takes place. So it is necessary to carry out such summation in the obtained cross section. To do so we use a standard R-matrix parametrization for multi-level S-matrix [10].

$$S_J(lj \to F\kappa\Pi c_f) = 2ie^{i\zeta_n^{lj}}(P_n^{lj})^{1/2} \sum_{\mu\nu} \gamma_{\mu n}^{lj} A_{\nu\mu}^J \gamma_{\nu f}^{JF\kappa\Pi c_f} (P_f^{c_f})^{1/2} e^{i(\omega_f^{c_f} + \zeta_f^{c_f})}. \qquad (5)$$

Only fission channel characteristics that enter into S-matrix are the amplitudes of partial fission widths. These are defined at scission point as overlap integral of FF channel function and a compound state WF

$$X_{vf}^{JM}(\omega,\beta,\tau) = \sum_{\Pi} \sum_{K>0} a_{vf}^{J\Pi K} \Psi_f^{J\Pi KM}(\omega,\beta,\tau) \qquad (6)$$

Here the random coefficients $a_{vf}^{J\Pi K}$ are measure of admixture of fission transition state $\Psi_f^{J\Pi KM}(\omega,\beta,\tau)$ to the compound-state v. Due to strong Coriolis mixing of different K components in WF of compound-states [11] these coefficients fluctuate over neutron resonances causing respective fluctuations of partial fission widths.

To calculate the partial fission amplitudes it is necessary at scission point to expand the internal part of TS WF over full set of 'future fission fragments' [6]

$$\Phi_m^K = \sum_{F'c_f'} u_{F'Km}^{c_f'} \sum_{K_1'K_1'} C_{J_1'K_1'J_2'K_2'}^{F'K} \chi_{J_1'K_1'}^{c_{f1}'} \chi_{J_2'K_2'}^{c_{f2}'}, \quad (7)$$

where the coefficients $u_{F'Km}^{c_f'}$ play role of FF pre-formation factors. In frame of R-matrix approach a calculation of the following overlap integral is carried out on radius a_f of fission channel c_f

$$\gamma_{vf}^{JF\kappa\Pi c_f} = \left(\frac{\hbar^2 a_f}{2\mu_f}\right)^{1/2} \int \varphi_{F|\kappa|\Pi JM}^{c_f}(n_f) X_{vf}^{JM}(\omega,\beta,\tau) d\Omega_{n_f} d\tau_f \quad (8)$$

Accounting for the explicit expressions of functions entering into the integrand and assuming that the orientations of deformation (ω) and fission (n_f) axes are coincided at scission it is possible to obtain [6] the following result ($K > 0$, $\Pi = \pi_1\pi_2(-1)^{J-F}$):

$$\gamma_{vfm}^{JF K\Pi c_f} = (\hbar^2 a_f / 2\mu_f)^{1/2} a_{vf}^{J\Pi K} \alpha_m^K \tilde{u}_{FKm}^{c_f} i^{(1-\Pi)(1-\pi_1\pi_2)/2}, \quad (9)$$

where coefficients $\tilde{u}_{FKm}^{c_f}$ defined [6] as $u_{FK}^{c_f} = i^{(1-\pi_1\pi_2)/2} \tilde{u}_{FK}^{c_f}$ are real. The obtained fission amplitudes are real too.

In above calculations it is assumed inexplicitly that helicity quantum numbers K_1, K_2 and κ are good just at scission. It is true for FF orbital momentum $L=0$. In general a helicity operator (Fn_f) does not commute with the centrifugal term of the Hamiltonian. So for $L>0$ helicity is not conserves. But recently discovered so called ROT effect [11] provides direct information on value of difference $\Delta\omega = \omega - n_f$. It is rather small $\Delta\omega \sim 10^{-3}$ so with near the same accuracy helicity has to conserve just at scission (see additional arguments in ref. [5]).

The differential cross-section defined for FF channel c_f includes summation of the bilinear combinations of S-matrix elements or respectively partial fission amplitudes $\gamma_{vf}^{JF\kappa\Pi c_f}$.

It is possible to carry out summation over c_f by substitution of the above expression for partial fission amplitudes. The result takes a form [6]:

$$\sum_{c_f}\sum_F i^{(\Pi'-\Pi)\pi_1\pi_2/2} \gamma_{vf}^{J'FK\Pi c_f} \gamma_{vf}^{JFK\Pi c_f} P_f^{c_f} = i^{(\Pi'-\Pi)/2} \gamma_{vf}^{J'\Pi K} \gamma_{vf}^{J\Pi K}, \qquad (10)$$

In which the reduced fission amplitudes $\gamma_{vf}^{J\Pi K}$ is introduced [6]:

$$\gamma_{vf}^{J\Pi K} = a_{vf}^{J\Pi K} v_f^K, \qquad v_f^K = \left(\sum_{c_f} (\alpha_m^K)^2 \sum_F \frac{\hbar^2 a_f}{2\mu_f} (\tilde{u}_{FK}^{c_f})^2 P_f^{c_f} \right)^{1/2} \qquad (11)$$

It is seen that summation over c_f and F results in no loss of inter-resonance interference. The reduced fission amplitudes $\gamma_{vf}^{J\Pi K}$ are defined for the effective fission channels JΠK and consist of random factors $a_{vf}^{J\Pi K}$ related to compound state WF v and the factors v_f^K which include dynamical characteristics of fission and that are the same for all resonances v.

The observed differential cross-section of (n,f,)-reaction

$$\frac{d\sigma_{nf}}{d\Omega_f} = \sum_{c_f} \frac{d\sigma^{c_f}}{d\Omega_f} =$$

$$= \pi\lambda^2 \sum_{J'J} (g_{J'}g_J)^{1/2} \sum_{l'j'lj} \sum_Q \Phi_{l'j'ljJ'J}^Q (n_f n_k n_f n_s) B_Q(l'j'lj;J'J) \qquad (12)$$

is expressed now in terms of the modified coefficients

$$B_Q(l'j'lj;J'J) = \sum_{c_f} B_Q^{c_f}(l'j'lj;J'J) =$$

$$= \sum_{\Pi'\Pi} \frac{1+\Pi'\Pi(-1)^Q}{2} i^{(\Pi'-\Pi)/2} \sum_{K\geq 0} C_{JKQ0}^{J'K} S_{J'}^*(l'j' \to K\Pi f) S_J(lj \to K\Pi f) \qquad (13)$$

that include the reduced S-matrix

$$S_J(lj \to K\Pi f) = 2ie^{i\zeta_n^{lj}}(P_n^{lj})^{1/2} \sum_{\mu\nu} \gamma_{\mu\nu}^{J\Pi j} A_{\mu\nu}^J \gamma_{vf}^{JK\Pi} \qquad (14)$$

So instead of unobservable cross-section defined for particular FF channel we obtained real differential (n,f) cross-section describing in principle all characteristics of low energy nuclear fission. It is appropriate to note that these formulas give consistent substantiation of the Reich-Moore approximation proposed [10] originally for description total fission cross-section and moreover extend this approach for the differential cross section. The formulas (11-14) give consistent extension of the original approach [8] for description of angular distribution of FF.

3. Analysis of experiments

For resonance neutron induced fission P-even and P-odd angular correlations of FF could be described by interference only of s- and p-wave (n,f) amplitudes as shown in Fig.2.

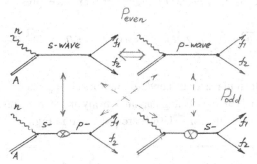

Fig. 2. The amplitudes of (n,f)-reaction contributing to P-even and P-odd angular correlations of FF.

Using formulas (12-13) it is easy to obtain explicit expression describing P-even FF angular correlations presented in Fig. 2.

$$\frac{d\sigma_{nf}(E)}{d\Omega_f}(1/4\pi)(\sigma_{nf}^0(E)+\sigma_{nf}^{FB}(E)(n_f n_k)+\sigma_{nf}^{RL}(E)p_n(n_f[n_f n_s])+$$

$$+\sigma_{nf}^{(2)}(E)f_2 P_2(n_f n_k))) \tag{15}$$

Here total fission cross section has a standard form

$$\sigma_{nf}^0(E)=\pi\lambda^2\sum_J g_J\sum_{K>0}\left|S_J(E;0\frac{1}{2}\rightarrow K\Pi)\right|^2 \tag{16}$$

and forward-backward correlation is written as:

$$\sigma_{nf}^{FB} = \pi \lambda^2 \sum_{J'Jj} \sum_{K \geq 0} q(J'JjK) Im[S_{J'}^*(1j \to K - \Pi f) S_J(0\frac{1}{2} \to K\Pi f)] \qquad (17)$$

The right-left correlations depend on value of neutron polarization p_n and have following expression:

$$\sigma_{nf}^{LR} = \pi \lambda^2 \sum_{J'Jj} (-\xi_j) \sum_{K \geq 0} q(J'JjK) Re[S_{J'}^*(1j \to K - \Pi f) S_J(0\frac{1}{2} \to K\Pi f)] \qquad (18)$$

Here $q(J'JjK) = \prod_0 g_J (-1)^{3/2-j} [6(2j+1)]^{1/2} U(IjJ1; J'\frac{1}{2}) C_{JK10}^{J'K}$ and $\xi_{1/2}=1$ and $\xi_{3/2}= -0.5$.

The angular anisotropy of FF emitted from neutron induced fission of aligned target nuclei depends on a degree of alignment f_2 and is expressed in the form:

$$\sigma_{nf}^{(2)} = \pi \lambda^2 G \sum_{J^\pi} \sum_{J'j} \sum_K g_J U(\frac{1}{2} IJ'2, JI) C_{JK20}^{J'K} S_{J'}^*(0\frac{1}{2} E \to K\Pi f) S_J(0\frac{1}{2} E \to K\Pi f) \qquad (19)$$

Using this last formula and a standard multilevel approach for S-matrix the experimental data [13] on FF angular anisotropy in s-wave (n,f)-reactions had been analyzed. The results for ^{235}U(n,f) are presented in Fig. 3.

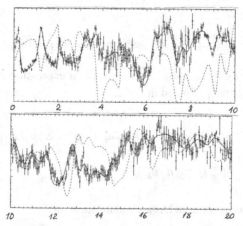

Fig. 3. The dependence of angular anisotropy A_2 of FF emitted from ^{235}U(n,f)-reaction (presented in arbitrary units) on neutron energy (eV). Points with error bars- experiment [13], solid curve- fit of $A_2(E)$ data together with all known cross sections, dashed curve- calculation of $A_2(E)$ with the resonance parameters describing only total fission cross section.

The value $A_2(E) = \sigma_{nf}^{(2)}(E)/\sigma_{nf}^{0}(E)$ shows the distinct inter-resonance interference and is very sensitive to choice of resonance parameters (compare solid and dashed curves). The fit shown by solid curve reproduce not only $A_2(E)$ dependence but as well it reproduces all relevant experimental cross sections known for this neutron energy range. Dashed curve shows the calculation of $A_2(E)$ with resonance parameters from ENDFB6. Partial fission widths $\gamma_{vf}^{J\Pi K}$ for compound states with J=3⁻ and 4⁻ strongly fluctuate and obey Porter-Tomas distributions with degree freedom equal unity. These fluctuations confirm a validity of results [10] on strong Coriolis mixing of different K components of compound-state WF. It was shown [7] that the fission channel with JΠK=4⁻0 is not totally forbidden by parity contrary to that it was assumed in previous considerations. As shown in ref. [13] both channels JΠK=3⁻0 and 4⁻0 are hindered in comparison with K≠ 0 channels. It could be connected with dependence of the parameters of two-hump fission barriers on additional quantum numbers related to different possible symmetries of TS WF (1). There are **R**-symmetry with quantum number $r = \pm 1$ and $S = PR$ one with respective quantum number $s = \pm 1$. The numbers r and s realize a dependence of fission barrier characteristics on total spin and parity of system at fixed value of K projection. Indeed for the 3⁻0 channel at the first fission barrier r = -1 (hindred) and for the second one s = + 1 (allowed). For 4⁻0 case a situation is just opposite - at first barrier r = + 1 (allowed) and at second one s = - 1(hindred).

More complicated type of FF angular correlations involving interference of s-wave and p-wave amplitudes of (n,f)-reaction could be analysed with aid formulas (17-18). The results for ^{239}Pu(n.f) is shown in Fig.4.

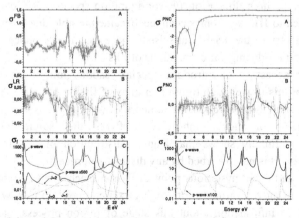

Fig.4. P-even (left) and P-odd (right) angular correlations of fission fragments emitting from ^{239}Pu(n.f)-reaction in dependence on neutron energy. Points with error bars show experimental data and solid curves present the fit [14].

As only $K=0$ channels contribute in ^{239}Pu(n.f)-reaction so for analysis of P-odd correlations shown in right side of Fig.4 it was used the following formula

$$\sigma^{PV} = \pi\lambda^2\sqrt{3}p_n(n_f n_k)Im\begin{bmatrix}S_1^*(0\frac{1}{2}\to 0\Pi\!f\,)S_0^{PV}(0\frac{1}{2}\to 0-\Pi\!f\,)- \\ +S_1^{*PV}(0\frac{1}{2}\to 0-\Pi\!f\,)S_0(0\frac{1}{2}\to 0\Pi\!f\,)\end{bmatrix} \quad (20)$$

Following ref. [12] we assumed that space parity violation (PV) takes place on stage of compound nucleus. So the PV effect results in appearance of small additional partial fission amplitudes for s-wave compound-state leading to p-wave fission channel.

In the end of this consideration it is necessary to note that the first idea how it is possible to describe P-odd and P-even FF angular correlations was proposed in ref. 15. But authors of [15] did not derive a consistent theory of (n,f)- reaction and due to this they could not introduce inter-resonance interference in proper way and was forced to use the artificial complex phases for partial fission amplitudes. This did not allow them to realize consistent analysis of experimental data.

4. Concluding remarks

Some semi-phenomenological conclusions could be derived from above analysis related to basis properties of nuclear fission mechanism. These are:

1. Axial symmetry of fissioning system is preserved with good accuracy. Reduction of high dimension FF configuration space to the low dimension one related to ЈПК A. Bohr's channels became possible due to specific pear-like symmetry of the TS WF at scission necessary to reproduce consistently P-even and P-odd angular correlations of FFs.
2. It is not necessary to introduce special "saddle point transition states". Instead of this it is enough to use hierarchy of fission barriers depending for each fission mode m on quantum numbers J,Π,K, s and r.
3. Conservation of angular interference effects after summation over FF channels definitely shows that angular distribution of FFs is formed at scission and it is not disturbed by any dissipative processes.
4. An open and even mysterious point of fission mechanism is transition from primary strongly deformed FF to highly excited ones. Up to now there is no any reliable information about this stage of fission process, namely when

(during Coulomb acceleration or after this) it is happened and how the respective configuration spaces match each other?

5. It is important to note that all above consideration was related to primary FFs. But internal WFs of really detected FFs differ from the ones defined at scission due change of their deformations as well as due to emission of neutrons and gamma-quanta. As it follows from experiment all these transformations do not change essentially the coherent properties of partial FF amplitudes and the resulting angular distributions.

In any case we see that resonance neutron induced fission is sensitive tool for study of basic feature of fission process that did not exhaust possibilities of more deep insight in its nature.

Acknowledgements

I express sincere gratitude to my colleagues A. L. Barabanov, A. B. Popov, Yu. Kopatch, N. Gundorin, J. Kliman, L. Kozlovsky, D. Tambovtsev, N. Gonin, G. Petrov, V. Petrova, A. Gagarsky, I. Guseva and V. Sokolov together with whom most results presented in this paper have been obtained.

References

1. N. Bohr and B. Mottelson, *Nuclear structure*, v.2, W. A. Benjaamin, Inc., NY, Amsterdam (1974).
2. D. L. Hill and J. A. Weeler, *Phys. Rev.*, **89, 577** (1953).
3. Yu. F. Smirnov and Yu. M. Chuvil'ski, *Phys. Lett.* B **134**, 25 (1984).
4. A. L. Barabanov and W. I. Furman, *Proc. of Int. Conf. on Nucl. Data for Science and Technology*, (Gatlinburg, Tennessee,1994), Ed. J.K. Dickens, **v.1, 448** (1994).
5. A. L. Barabanov and W. I. Furman, *Z. Phys.* A **357, 411** (1997).
6. A. L. Barabanov and W. I. Furman, *Czechoslovak journal of Physics, Suppl.* B **53**, 359 (2003).
7. A. L. Barabanov and W. I. Furman, *Phys. At. Nucl.* **72, 1259** (2009).
8. V. M. Strutinsky, *ZETPh* **30, 606** (1956).
9. W. Furman and J. Kliman, *Proc. 17th Int. Symp. on Nucl. Phys.* Eds. D. Seeliger and H. Kalka, ZfK, Drezden, **142** (1988).
10. S. G. Kadmensky, V. M. Markushev and W. I. Furman, *Phys. At. Nucl.* **35, 300** (1982).
11. F. Goennenwein, V. Bunakov, O. Dorvaux et al.,*Proc. of Seminar on fission*, Eds. C. Wagemans, J. Wagemans and P. D'hondt, WS,3, (2008).
12. V. E. Bunakov, W. I. Furman, S. G. Kadmensky et al., *Phys. At. Nucl.* **49, 988** (1991).

13. Yu. N. Kopach, A. B. Popov, W. I. Furman *et al, Phys. At. Nucl.* **62, 929** (1999).
14. A. L. Barabanov, W. I. Furman, A. B. Popov et al, *Proc. of 11-th Int. Seminar on Interaction of Neutrons with Nuclei*, Dubna, JINR E3-2004, **304** (2004).
15. O. P. Sushkov and V. V. Flambaum, *Sov. Phys. Usp.*, **136, 3** (1983).

PHENOMENOLOGICAL MODEL IMPROVEMENT FOR THE MONTE CARLO SIMULATION OF THE FISSION FRAGMENT EVAPORATION

O. LITAIZE* and O. SEROT

CEA Cadarache, F-13108 Saint-Paul-lez-Durance, France
* E-mail: olivier.litaize@cea.fr

The Monte Carlo simulation of the fission fragment evaporation is improved compared to our previous works in order to beter reproduce the prompt neutron spectra and gamma energy without perturbing other results on post-fission observables. The model is based on a mass dependent temperature ratio between primary fragments and a spin dependent excitation energy limit for neutron emission. In addition the total excitation energy is no longer entirely partitioned through an aT^2 relation due to fission fragment rotational energy.

1. Introduction

The Monte Carlo simulation of fission fragment (FF) has already been undertaken by different authors with more or less success. The basic ideas have been already developed by Lemaire et al.[1] and very recently by Randrup et al.[2] In Lemaire's work two kind of hypothesis related to the partitioning of the FF initial excitation energy at the scission point are considered. The first one is an equipartition of the temperature between the two complementary fragments. Under this hypothesis, the authors have shown that the average prompt neutron multiplicity as a function of mass can not be reproduced. The second one uses experimental results like mean neutron energy or average number of prompt neutrons to infer the initial excitation energy. Here, a nice agreement between experimental observables and calculations could be achieved, but this approach has no predictive power. In the present work various additional models have been developed and different hypothesis have been used (mass dependent temperature ratio of the fully accelerated complementary fragments, spin dependent excitation energy limit for neutron emission, ...) to improve the agreement with experimental data. In addition, a slight modification was performed compared to our previous work[3] in order to partition in a different way the excitation energy between complementary fragments.

2. Fission fragment evaporation model

The description of the model used in this work can be found elsewhere[3] but the main ideas are reminded hereafter.

The input experimental data are the fission fragment mass number A and the kinetic energy KE distributions measured by Varapai[4] for ^{252}Cf(sf). The nuclear charge Z necessary to select a nucleus is sampled from an almost Gaussian function defined by the most probable charge coming from unchanged charge density assumption corrected by an oscillating polarization function depending on the mass number.[5] Once the light FF is selected the heavy complementary one is deduced (neglecting ternary fission) and its kinetic energy sampled. Total excitation energy is then calculated and partitioned between fragments considering a mass dependent temperature ratio $R_T = T_L/T_H$. The residual nuclear temperature is caluated from level density parameter $a(A-1, Z)$, excitation energy $E^*(A-1, Z)$ and neutron separation energy S_n and is used in a Weisskopf spectrum for neutron evaporation.[6] The excitation energy after a neutron emission is then $E^*(A, Z) - S_n - \epsilon$ where ϵ is the centre of mass neutron energy. This procedure is applied up to a spin dependent excitation energy limit $E^*_{lim}(J) = S_n + E^{rot}(J)$. The rotational energy depends on the total angular momentum and the moment of inertia of the FF. Under this limit the excitation energy is used for gamma-rays emission (not yet implemented).

In this work we have improved the excitation energy partitioning between complementary fragments. In Ref.[3] we have considered that the total excitation energy TXE after fully acceleration was partitioned through an aT^2 relation (Fermi gas). Under this assumption an overestimation of the residual excitation energy available for gamma-rays after neutron emission was found. Now, we consider a more realistic hypothesis where TXE is a sum of an intrinsic excitation energy and a rotational energy. Then only the intrinsic energy part coressponding to $TXE - E^{rot}_L - E^{rot}_H$ is partitioned between the fragments. Under this assumption the prompt gamma energy as a function of mass is clearly better reproduced. In addition, the prompt neutron spectrum is also more consistent with reference data. The other post-fission observable distributions are still in good agreement with experimental data considering that the moment of inertia involved in the rotational energy is equal to 50% of a rigid body (instead of 40% in the previous reference) and the average total angular momenta are respectively

6ℏ and 7.2ℏ for light and heavy primary fission fragments (compared to 8 and 9 in the previous work).

In the next section, we compare the average prompt neutron multiplicity as a function of mass for different models. The following sections are devoted to prompt neutron spectrum, multiplicity distribution and prompt gamma energy calculated with the refined model.

3. Average prompt fission neutron multiplicity

The number of evaporated neutrons is mainly related to the FF nuclear temperature and the energy limit for the neutron evaporation process. We have reported in Table 1 the results from previous models (depending on the temperature ratio and the excitation energy limit) and the refined final model described in this work.

If we consider S_n for the excitation energy limit, the total average prompt neutron multiplicity is overestimated compared to Vorobyev's experimental data.[7] By considering an equipartition of the temperature between the two complementary fragments one can not observe the saw-tooth shape and the $\bar{\nu}_L/\bar{\nu}_H$ ratio is wrong as already mentioned by Talou.[8] By considering a temperature higher for the light FF than for the heavy FF,[9] a kind of saw-tooth shape appears but as said before the total average prompt neutron multiplicity is still overestimated. If we consider a spin dependent excitation energy limit then the moment of inertia involved in the rotational energy has to be about 50% of a rigid ellipsoid in order to reproduce the total average $\bar{\nu}$ but the $\bar{\nu}(A)$ is not satisfactory. A moment of inertia of a irrotational flow gives completely wrong results and a pure rigid body is not entirely satisfactory.

Finally the best agreement with experiment performed by Vorobyev et al.[7] is achieved if we consider a mass dependent linear law for the temperature ratio $R_T \equiv R_T(A)$, a spin dependent energy limit $S_n(J) = S_n + E^{rot}(J)$ for the neutron evaporation and a moment of inertia equal to 50% of a rigid body moment of inertia.

Some constraints based on nucleus shapes and shell closures are considered for the determination of the $R_T(A)$ temperature ratio law. For mass split 78/174 the light FF is near spherical and then its temperature is lower than its heavy partner leading to a minimum $R_T < 1$. For mass split 120/132 the situation is reversed and the heavy FF is near spherical with a higher temperature than its light partner leading to a maximaum $R_T > 1$. Finally for symetric mass split 126/126 the temperature is the same: $R_T = 1$. Bewteen these three configurations a linear law is assumed.

Table 1. Prompt neutron multiplicities and mean excitation energies for various models compared with Vorobyev's data.[7]

R_T	E^*_{limit}	$\mathcal{J}$	$\bar{\nu}_L$	$\bar{\nu}_H$	$\bar{\nu}$	$<E^*>_L$ (MeV)	$<E^*>_H$ (MeV)
1.	S_n	-	1.82	2.44	4.26	15.84	18.26
1.25	S_n	-	2.28	1.93	4.21	19.75	14.44
1.25	$S_n(J)$	$\mathcal{J}_{rig}$	2.18	1.83	4.01	19.74	14.42
1.25	$S_n(J)$	$0.5\mathcal{J}_{rig}$	2.07	1.71	3.78	19.74	14.41
$R_T(A)$	$S_n(J)$	$0.5\mathcal{J}_{rig}$	2.06	1.70	3.76	19.90	14.28
Vorobyev et al. (2004)			2.051	1.698	3.756	-	-

The goal of this work is not to accurately reproduce the experimental data because the complete model is not yet established and some post fission observables not accurately measured. Nevertheless we can observe in Fig. 1 that the results calculated with the refined model are quite comparable with experimental data.

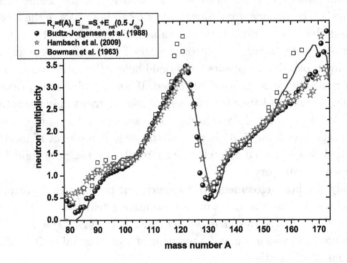

Fig. 1. Average prompt neutron multiplicity as a function of mass A compared with measurements from Budtz-Jørgensen,[10] Hambsch[11] and Bowman.[12]

4. Prompt neutron multiplicity distribution

The prompt neutron multiplicity distibution $P(\nu)$ is shown in Fig. 2. The comparison with experimental results from Ref.[7] shows an overall good

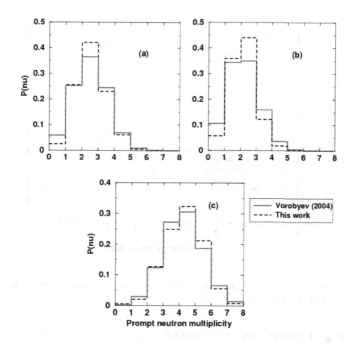

Fig. 2. Prompt neutron multiplicity distributions for ligth FF (a), heavy FF (b) and for both of them (c), compared with Vorobyev's data.[7]

agreement. These distributions are quite sensitive to the amount of angular momentum carried away by neutrons, which was fixed in this work to $1\hbar$ per evaporated neutron. We have pointed out that if we change this quantity the multiplicity distribution can be improved but other distributions are deteriorated highlighting the importance of the amount of angular momentum carried away by neutrons.

5. Prompt fission neutron spectrum

The prompt fission neutron spectrum PFNS in the laboratory frame is compared in Fig. 3 with the Manhart's evaluation[13] and a Maxwellian spectrum ($T = 1.42$ MeV) leading to a mean energy of 2.13 MeV. Here a value of 2.14 MeV was found. The discrepancy between our calculation and the Manhart's evaluation is under 2% from 500 keV to around 6 MeV. Some way of improvement can be the accounting for an energy dependent cross section for the inverse process of compound nucleus formation in the Weisskopf spectrum as already reported by Madland and Nix in Ref.[14]

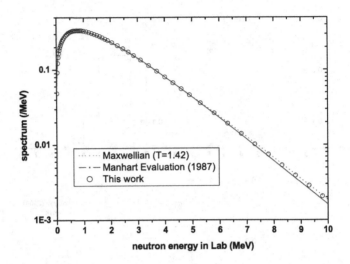

Fig. 3. Neutron spectrum obtained with our model in comparison with a Maxwellian ($T_M = 1.42$) and the Manhart evaluation.[13]

6. Average prompt gamma energy

Even if the gamma-rays deexcitation is not yet implemented, the average fission fragment excitation energy leading to prompt gamma emission can be deduced for each secondary fragment: $<E_\gamma>(A)$. Using the new assumption described in section, we have now reduced the discrepancy observed in Ref.[3] between calculation and measurement (see Fig. 4).

7. Conclusion

As shown in this paper, various parameters can strongly influence the prompt neutron characteristics. The most important parameters are the repartition of the excitation energy between the complementary fragments, their moment of inertia and initial spin distributions and lastly the excitation energy limit for neutron emission. The problem is obviously multidimensional: the so called saw-tooth can be well reproduced with a given set of parameters, the other distributions being distorted. Nevertheless, under the models and assumptions discussed in this paper, the main features of the prompt neutrons as well as the excitation energy available for prompt γ-emission are nicely reproduced.

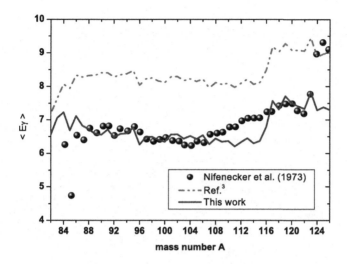

Fig. 4. Excitation energy available for γ deexcitation $< E_\gamma > (A_L + A_H)$ as a function of the mass number of the light fragment compared with results reported by Nifenecker.[15]

References

1. S. Lemaire et al., *Phys. Rev.*, **C72**, 054608(2005).
2. J. Randrup and R. Vogt, *Phys. Rev.*, **C80**, 024601(2009).
3. O. Litaize and O. Serot, *Proc. Int. Conf. on Nuclear Data for Science and Technology ND2010*, Jeju Island, Korea, April 26-30, 2010.
4. N. Varapai et al., *Proc. Int. Workshop on Nuclear Fission and Fission Product Spectroscopy*, Cadarache, France, May 11-14, 2005.
5. A.C. Wahl et al., *Phys. Rev.*, **C126**, 1112(1962).
6. V. Weisskopf, *Phys. Rev.*, **52**, 295(1937).
7. A.S. Vorobyev et al., *Proc. Int. Conf. on Nuclear Data for Science and Technology ND2004*, Santa Fe, USA, Sept. 26-Oct. 1, 2004.
8. P. Talou, *Proc. Int. Workshop on Nuclear Fission and Fission Product Spectroscopy*, Cadarache, France, May 13-16, 2009.
9. T. Ohsawa, Report INDC-NDS-251 (1991)
10. C. Budtz-Jørgensen et al., *Nucl. Phys.*, **A490**, 307(1988).
11. F.-J. Hambsch et al., *Proc. Int. Workshop on Compound Nuclear Reactions and Related Topics (CNR2009)*, Bordeaux, France, Oct. 5-8, 2009.
12. H. R. Bowman, *Phys. Rev.*, **129**, 2133(1963).
13. W. Manhart, *Report IAEA-TECDOC*, **410**, 158(1987).
14. D.G. Madland and J.R. Nix, *Nucl. Sci. Eng.*, **81**, 213(1982).
15. H. Nifenecker et al., *Proc. Int. Conf. on Physics and Chemistry of Fission*, Rochester, New-York, USA, Aug. 13-17, 1973.

EXCITATION ENERGY SORTING IN SUPERFLUID FISSION DYNAMICS

B. JURADO and K.-H. SCHMIDT

CENBG, CNRS/IN2P3, Chemin du Solarium B.P. 120, 33175 Gradignan, France

It is now well established that at moderate excitation energies the nucleus temperature does not vary with increasing excitation energy. We show that, as a consequence, two nuclei with different temperatures brought into contact show a rather surprising energy-sorting mechanism where the hotter nucleus transfers all its excitation energy to the colder one. The scission configuration of the fission process offers a unique possibility to observe this phenomenon. The energy sorting mechanism is clearly reflected by the mean number of prompt neutrons as a function of the fragment mass and by the dependence of the local even-odd effect with mass asymmetry.

Keywords: Level density; excitation energy sharing in fission; prompt neutron yields; even-odd effect.

1. Introduction

Experimental data on nuclear level densities[1] show that the behavior of nuclei at moderate excitation energies E^* is well described by a constant temperature level density of the form:

$$\rho(E^*) \propto exp(\frac{E^*}{T}). \tag{1}$$

where T is the constant temperature parameter. This means that in this regime the temperature of nuclei does not vary with E^*. The main reason for this constant-temperature behavior is that pairing correlations lead to an effective number of degrees of freedom that increases in proportion to E^*. Cooper pairs of neutrons and protons melt in a way that the mean energy per nucleonic excitation and thus the nuclear temperature stays constant. In a recent experimental work on level densities[2] it has been surprisingly found that the constant temperature behavior remains up to 20 MeV E^*. T. v. Egidy et al.[1] have fitted expression (1) to experimental data on level densities for nuclei ranging from ^{18}F to ^{251}Cf. The following dependence of the constant temperature T with the mass A of the nucleus and the shell correction S was found:

$$T = A^{\frac{-2}{3}}(17.45 - 0.51S + 0.051S^2). \qquad (2)$$

In this work we investigate how two nuclei in the particular regime of constant temperature behave when they are set in thermal contact. This situation is realized near the scission configuration in fission.

2. Excitation energy sorting

In fission, the energy difference between the ground-state masses of the initial fissioning system and the final fission fragments, given by the Q value, and the initial excitation energy of the fissioning nucleus E^*_{CN}, ends up either in the total excitation energy (TXE) or in the total kinetic energy (TKE) of the fragments. In the present work, we consider low-energy fission with initial excitation energies E^*_{CN} up to a few MeV. Therefore, the TXE is released by neutron evaporation and gamma emission from the fission fragments. We may distinguish three classes of energy at scission, which add up to the final TXE of the fission fragments: (i) Intrinsic excitations by single-particle or quasi-particle excitations. (ii) Deformation energy. (iii) Collective excitations stored in normal modes. The intrinsic excitation energy is the sum of the E^* above the barrier height and the fraction of the difference in potential energy between saddle and scission that is dissipated into intrinsic excitations. The deformation energy ends up as part of the E^* available when the fission fragments recover their ground-state deformations. The damping of collective excitations as for example bending and wriggling modes leads to additional E^* in the fragments.

We assume that already somewhat before the scission configuration the two nascent fragments have acquired their individual properties concerning shell effects and pairing correlations and can be treated as two well defined nuclei set in thermal contact through the neck. The division of intrinsic excitations can be derived when thermal equilibrium is assumed among the intrinsic degrees of freedom in each fragment. As said above, the nuclear level density at low E^* is very well described by the constant-temperature formula (1) with a specific value of T for each fragment. This leads to a very interesting situation for the two nascent fragments near the scission-point configuration: There is no solution for the division of intrinsic E^* with T1=T2. As long as some excitation energy remains in the fragment with the higher temperature, its E^* is transferred to the fragment with the lower temperature. That means, a process of E^* sorting takes place where all E^* accumulates in the fragment with the lower value of the T parameter, while the other fragment looses its entire E^*. According to eq. (2) the heavy fragment generally has the lower T and thus attracts all the E^*. This behaviour is unique. To our knowledge all other systems in nature reach thermal equilibrium with T1=T2 before the thermal energy of the hotter object is completely exhausted. The flow of E^* from the hot fragment to the cold fragment can be seen as a way to maximize the number of occupied states or its entropy. The number of available states of the light nucleus is small compared to that of the complementary fragment. Therefore, the situation in which the light nucleus has part of the E^* leads to a smaller entropy than the situation in which the entire E^* is transferred to the heavy nucleus which has considerable more available states. We have thoroughly described the energy sorting process in Ref.[3] It is also one of the ingredients of the GEF code.[4]

3. Neutron yields as a function of the fragment mass

It is of great interest to find experimental signatures that reflect the excitation energy sorting. These signatures must be able to distinguish between the intrinsic E^* accumulated in each fission fragment. The number of evaporated neutrons as a function of the fragment mass is directly proportional to the E^* of the fragment - except an offset of a few MeV that is taken away by gamma emission- and therefore should clearly reflect the peculiar situation of the full transfer of the intrinsic E^* to the cold fragment. The neutron-induced fission of ^{237}Np has been studied very carefully at two different neutron energies.[5] Fig. 1 shows the average number of evaporated neutrons as a function of the fragment mass. The well known saw-tooth-like behavior of this curve is attributed to the deformation energy. For asymmetric mass splits we observe a very peculiar feature: Fig. 1 shows that the increase of E^* leads to an increase of the number of evaporated neutrons for the heavy fragment, only. Actually, a quantitative analysis of the data reveals that all of the increased E^* appears in the heavy fragment. This observation is rather general as it was also found for other fissioning systems such as ^{233}U and ^{238}U and other incident particles like protons.[6-9] However, no clear explanation has yet been found for this effect. Actually, this is a direct consequence of the different constant temperatures of the two fragments at scission. According to eq. 2, the temperature of the heavy fragment, in the absence of strong shell effects, is always lower than the temperature of the light fragment. Therefore, the heavy fragment will absorb the entire available intrinsic E^* and evaporate more neutrons. We would like to stress

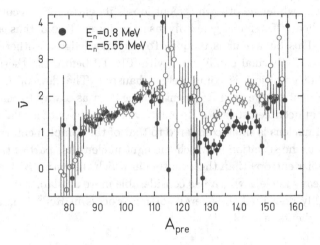

Fig. 1. Average number of prompt neutrons as a function of the primary fragment mass for the neutron-induced fission of ^{237}Np, data taken from ref.[5]

that our argumentation is based on the same assumptions as other work that investigates the sharing of intrinsic E^* at scission (see e.g.[10]). That is, we have assumed independent fission fragments and thermal equilibration between the fragments at scission. What is substantially different is that we use the constant-temperature level density which correctly describes the behaviour of nuclei at moderate E^* and not the commonly used Fermi gas level density[11] which is only valid at high E^*.

4. Even-odd effect in fission fragment element yields

At the end of the energy sorting process the heavy fragment can gain additional E^* if few nucleons are transferred through the neck so that the light fragment becomes even-even. The gain in E^* can be up to four times the pairing gap. Therefore, according to the energy sorting mechanism, there will be a tendency for the hot (normally the light) fragment to be fully paired. Let us now consider the dynamics of the energy sorting process. The time t to form a fully paired light fragment is the sum of the time needed for the light fragment to transfer all its E^* to the heavier one, which we will cal t_{E^*}, and the time to exchange few nucleons through the neck t_{exch}. Time t_{exch} is rather short so that the time t is dominated by t_{E^*}. The latter

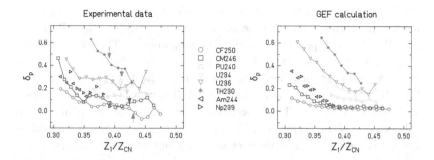

Fig. 2. Local even-odd effect as a function of asymmetry for various fissioning nuclei. The experimental data are shown on the left and the GEF calculation[4] on the right.

will increase with the initial excitation energy in the light fragment $E^*_{0,light}$ since it will take a longer time to transfer all the energy from the light to the heavy nucleus. We consider that the initial excitation energy $E^*_{0,light}$ is proportional to the available excitation energy at scission E^*_{sci} which, as was said above, is the sum of the excitation energy at saddle E^*_{CN} and the dissipated energy between saddle and scission $E^*_{sad-sci}$. E^*_{CN} increases with beam energy and $E^*_{sad-sci}$ with the Coulomb parameter $Z^2/A^{1/3}$ since the saddle to scission path becomes longer.[12] On the other hand, the time t will decrease when the temperature difference $T1 - T2$ between the two fragments increases. A higher temperature gradient leads to faster flow of E^* between the two fragments. According to eq. (2) an increase in temperature difference corresponds to an increase in the asymmetry of the mass split. To resume, the time t follows the expression:

$$t \propto \frac{E^*_{sci}}{T1 - T2}. \qquad (3)$$

As a consequence, t will increase with the beam energy and the mass of the fissioning nucleus and will decrease with increasing asymmetry of the mass split. Let us now assume that there exists a time t_p above which the exchange protons through the neck is very much hindered [a]. If $t > t_p$, no net even-odd effect is induced because protons cannot be transferred to the heavy nucleus. Thus, according to the energy sorting process the even-odd effect as a function of asymmetry should have a threshold character. The

---

[a]This hindrance is due to the Coulomb barrier in the neck region which increases with the distance between the two fragments. The difference between the inner barrier for neutrons and protons is shown in Fig. 16 of ref.[13] Although the latter calculations have been done for transfer reactions, we expect similar results for fission.

threshold asymmetry where the even-odd effect created by the energy sorting sets in (corresponding to the asymmetry for which $t = t_p$) will increase with the Coulomb parameter of the fissioning nucleus. The left part of Fig. 2 shows experimental data on the local even-odd effect as a function of charge asymmetry for different fissioning nuclei. The data have been taken from Ref.[14] It presents data measured at ILL where fission was induced with thermal neutrons. The experimental data illustrate clearly how the local even-odd effect decreases with increasing mass of the fissioning nucleus. For a fixed even-Z fissioning nucleus, the general trend presented by the data is a small and rather constant even-odd effect close to symmetry and a strong increase as we move to more asymmetric fission. The latter feature occurs at an asymmetry value that increases with the mass of the fissioning nucleus. For ^{230}Th this change is not shown by the data. However, we presume that this is because the threshold asymmetry for this nucleus is close to symmetry where no data have been measured. It is interesting to note that the even-odd effect of odd-Z fissioning systems at large asymmetry follows the general behavior of the even-Z fissioning systems. Their values are very close to those of the neighboring even-Z systems. For several systems, the data point in Fig. 2 that is closest to symmetry is appreciably higher than expected from the global trend. This effect may be associated to the influence of the Z=50 shell in the complementary fragment, which is known to enhance the yield of tin isotopes and, thus, leads to a local increase of the deduced even-odd effect. In the GEF code[4] the dependence of the local even-odd effect with asymmetry is modeled with a Gauss-integral. The threshold asymmetry value (the value at which the Gauss-integral is 1/2) is the one that full fills the condition $C * E^*_{sci}/|T1 - T2| = t_p$, where C is a constant. t_p/C is adjusted to the data and has the same value for all nuclei. The width of the Gauss function is set proportional to $|T1-T2|$. The scaling factor for $|T1-T2|$ is the same for all nuclei and fitted to the experimental data. On the right part of Fig. 2 the results of our model GEF for the same fissioning systems are presented. All the features mentioned above are nicely reproduced by our calculation. The energy sorting mechanism also predicts that, for a given fissioning nucleus, the threshold asymmetry should increase with increasing beam energy. Unfortunately there are no data to verify this statement.

5. Conclusions

Nuclei at low E^* are peculiar systems where the temperature remains constant with increasing E^*. The very special feature of this phenomenon in nuclei is that the constant-temperature regime reaches down to zero energy. The scission configuration of the fission process offers the unique possibility to investigate how two different nuclei in this special regime of constant temperature share the available intrinsic excitation when they are in thermal contact. We have shown for the first time that the excitation energy keeps flowing from the hot to the cold fragment until the excitation energy of the hot fragment is completely exhausted. The E^*-sorting effect explains very easily why an increase in E^* leads to an increase of the number of prompt neutrons of the heavy fragment, only. Indeed, the temperature of the heavy fission fragments is generally lower than that of the light ones. Therefore, all the intrinsic E^* is cumulated in the heavy fragment. The energy sorting process is also reflected by the dependence of the even-odd effect in fission fragment element yields with asymmetry. This dependence has a threshold character with a threshold asymmetry that increases with beam energy and the Coulomb parameter of the fissioning nucleus.

Acknowledgments

This work was supported by the EURATOM 6. Framework Programme "European Facilities for Nuclear Data Measurements" (EFNUDAT), contract number FP6-036434. We thank F. Kaeppeler for providing us with the numerical values of the neutron-yield data.

References

1. T. von Egidy and D. Bucurescu, *Phys. Rev.* **C 72**, 044311 (2005).
2. A.V. Voinov et al., *Phys. Rev.* **C 79**, 031301(R) (2009).
3. K.-H. Schmidt and B. Jurado, *Phys. Rev. Lett.* **104**, 212501 (2010).
4. www.cenbg.in2p3.fr/GEF and K.-H. Schmidt, B. Jurado, these proceedings.
5. A. A. Naqvi et al., *Phys. Rev.* **C 34**, 212501 (2010).
6. S. C. Burnett et al., *Phys. Rev.* **C 3**, 2034 (1971).
7. C. J. Bishop et al., *Nucl. Phys.* **A 150**, 129 (1970).
8. R. Mueller et al., *Phys. Rev.* **C 29**, 885 (1984).
9. M. Strecker et al., *Phys. Rev.* **C 41**, 2172 (1990).
10. S. Lemaire et al., *Phys. Rev. C* **72**, 024601 (2005).
11. H. A. Bethe, *Phys. Rev.* **50**, 332 (1936).
12. M. Asghar, R.W. Hasse, *J. Phys.* **C 6**, 455 (1984).
13. W. von Oertzen and A. Vitturi, *Rep. Prog. Phys.* **64**, (2001) 1247.
14. F. Rejmund et al., GANIL P 2010-01, http://hal.in2p3.fr/in2p3-00487186/fr/.

SYSTEMATIC ANALYSIS OF STRUCTURAL EFFECTS IN FISSION-PRODUCT YIELDS AND NEUTRON DATA AND THE CONSEQUENCES FOR OUR UNDERSTANDING OF THE FISSION PROCESS AND THE PREDICTIVE POWER OF MODEL PREDICTIONS

K.-H. SCHMIDT and B. JURADO

CENBG, CNRS/IN2P3, Chemin du Solarium B.P. 120, 33175 Gradignan, France

Structural effects in fission-product yields and neutron data for a large number of fissioning nuclei between ^{220}Th and ^{256}Fm from spontaneous fission to 14-MeV-neutron-induced fission have been used to deduce information on the properties of the fissioning systems. Macroscopic properties are attributed to the compound nucleus, while fission channels are ascribed to shells in the nascent fragments. Using a recent general empirical description of the nuclear level density and assuming different characteristic time scales for the collective degrees of freedom of the fissioning system, a new fission model has been developed. The model combines the statistical concept of the scission-point model of Wilkins et al. with empirically determined properties of the potential-energy surface and some characteristic dynamical freeze-out times. Although no fine tuning of the parameters has yet been performed, the model reproduces all measured fission yields and neutron data rather well with a unique set and a relatively small number of free parameters. Since the parameters of the model are closely related to physical properties of the systems, some interesting conclusions on the fission process can be deduced. Prospects for the predictive power of this semi-empirical approach for hitherto unknown fissioning systems are discussed.

Keywords: Fission-fragment yields; fission channels; fission dynamics; macroscopic-microscopic approach; separability principle; energy sorting; even-odd effect.

1. Introduction

Ordering schemes, systematics and semi-empirical models are powerful approaches for advancing our understanding of complex phenomena in nature. In nuclear fission great progress has been made by introducing the general concept of fission channels.[1] It established a link between the observed characteristics, e.g. in fission yields and kinetic energies, and the properties of the potential-energy surface of the fissioning system. However, it did not allow for quantitative predictions. The theoretical description of nuclear fission, in particular at low excitation energies with its rich manifestation of nuclear-structure phenomena is still a challenge. At present, one is restricted to purely empirical models (e.g.[2]) for a good quantitative description of the data.

In contrast, the theoretical description of atomic masses has reached a high degree of precision, and the complex phenomena behind the manifold global and structural effects are quantitatively rather well understood. The data are very well reproduced by models based on the macroscopic-microscopic approach, while fully microscopic models are supposed to be more realistic for nuclei close to the drip lines.

In the present contribution we try to profit from the successful concepts and methods used in mass models in order to establish an improved model of the fission process. We make use of several well-known and a few newly developed concepts to develop a description for fission-fragment distributions and the properties of prompt neutrons, which reproduces the experimental data with high precision and which is expected to have a high predictive power for systems that have not been measured and that are not accessible to experiment.

2. Reminder on methods and concepts used in mass models

Atomic mass models[3] span the range from local formulas,[4] directly based on measured mass values, to microscopic models based on effective nucleon-nucleon interactions. However, intermediate approaches proved to be the most successful ones for a long period. A rather good description of the binding energy of atomic nuclei has been proposed by C. F. von Weizsaecker already in 1935. It relies on the analogy of an atomic nucleus with an electrically charged drop of a classical liquid. By additionally considering the Fermionic nature of the nucleons by the asymmetry term, the liquid-drop model reproduces the nuclear binding energies with a precision of about 1 per cent. In the macroscopic-microscopic approach, structural effects due to shell effects and pairing correlations are calculated separately by the Strutinsky method[5] and added to the value obtained by the liquid-drop model. The liquid-drop model still gives a very good estimation of the global behaviour of nuclear binding for nuclei not too close to the drip lines, which is at present hardly reached by microscopic models that rely on the interactions of nucleons governed by an effective nuclear force.

A systematic analysis of empirical data and a careful comparison with global models, like the liquid-drop model, have proven to be very useful in establishing evidence for phenomena, which go beyond the basic description. Exceptionally high binding of nuclei along "magic numbers" due to shell effects,[6,7] even-odd structure due to pairing correlations and the manifestation of the congruence energy[8] were recognized by systematic deviations from the liquid-drop predictions. The role and the magnitude of the spin-orbit force have been deduced,[9,10] and the appearance of new magic numbers far from stability has been evidenced.[11] The comparison of nuclear properties with a global background acts as a magnifying glass on structural effects and new phenomena and, thus, forms the important counterpart to microscopic models, which try to model the complex phenomena on a more fundamental level. One should not forget that also microscopic nuclear models remain phenomenological,[12] since the effective force is adjusted to reproduce best the body of experimental data.

3. Concept of a general fission model

The experimental information available in low-energy fission of a specific nucleus is by far more rich than just one numerical value like its atomic mass: These are the many individual nuclide yields, the kinetic energies of the fission fragments, the prompt neutron yields and neutron energies, to mention the most prominent ones, only. Moreover, the fission observables are the result of a complex dynamical process, while the ground state of a nucleus is the energy of an equilibrium state. Thus, the modeling of the fission process appears to be much more difficult.

Any fission model needs to follow the dynamic evolution of the fissioning system up to scission. The number of protons and neutrons in the two fragments, their kinetic energies, the available energies above their respective ground states as well as their angular momenta are decided or can uniquely be deduced from the scission configuration. However, it is not justified to assume statistical equilibrium at scission as it was done by Wilkins et al.,[13] because a considerable inertia may prevent the system to adjust instantaneously to the bottom of the potential-energy valley on the fission path. One may assume that there is a dynamical freeze-out somewhere before reaching scission, which is specific to the different collective variables. The mass asymmetry degree of freedom is characterized by a rather early freeze-out due to its large inertia, while the N/Z degree of freedom is decided later, because the mass transport and consequently the inertia associated with the charge polarization is much lower. Thus, there is no single, well defined configuration, where a statistical-model assumption seems to be justified.

Due to this difficulty and the unavoidable uncertainty of a theoretical fully dynamic calculation, we decided to extract the relevant information from the available experimental data directly. The measured characteristics of the distributions in the different variables contain the required information in the most precise and realistic way. However, it is not clear, whether this approach is feasable, because we should establish this empirical information for each fissioning system independently. Thus, this approach would be equivalent to a purely empirical model with a specific parameter set for each fissioning system. One cannot expect a high predictive power for unmeasured systems from this kind of approach.

The application of the separability principle[14] solves this problem. Indeed, two-centre shell-model calculations revealed that the shell effects of the fissioning system already immediately beyond the outer saddle are very similar to the sum of the shells in the two nascent fragments.[15] The combination of this finding with the macroscopic-microscopic approach leads to a very important conclusion: The shell effects on the fission path are associated to the nascent fragments. Essentially the same shell effects are present if the same fragments are formed in different fissioning systems. Thus, the full body of experimental data on fission-fragment properties can be used to deduce the relevant information on shell effects on the fission path, which are the same for all fissioning systems. Only the macroscopic potential on the fission path is specific to the fissioning system. The separability principle of microscopic effects, which are associated to the nascent fragments, and of macroscopic effects, which are specific to the fissioning system, make our approach feasable and gives it a high predictive power.

4. Formulation of the model

A basic ingredient of the model is the curvature of the macroscopic potential in mass-asymmetry on the fission path at freeze out of the asymmetry degree of freedom.[16] This value determines the width of the mass-symmetric fission channel and, even more importantly, the relative strengths of the asymmetric fission channels. The shell effects, which are responsible for the asymmetric fission valley are fully effective, if they appear close to symmetry. This is the case for the heavier actinies. In contrast, the influence of these shell effects is weakened in the lighter actinindes, where these shells appear at larger asymmetry. This interplay of the macroscopic potential and the shell effects determines the transition from single-humped mass distributions to double-humped distributions around $A = 226$.

We present here in some detail the analysis of the mean position of the

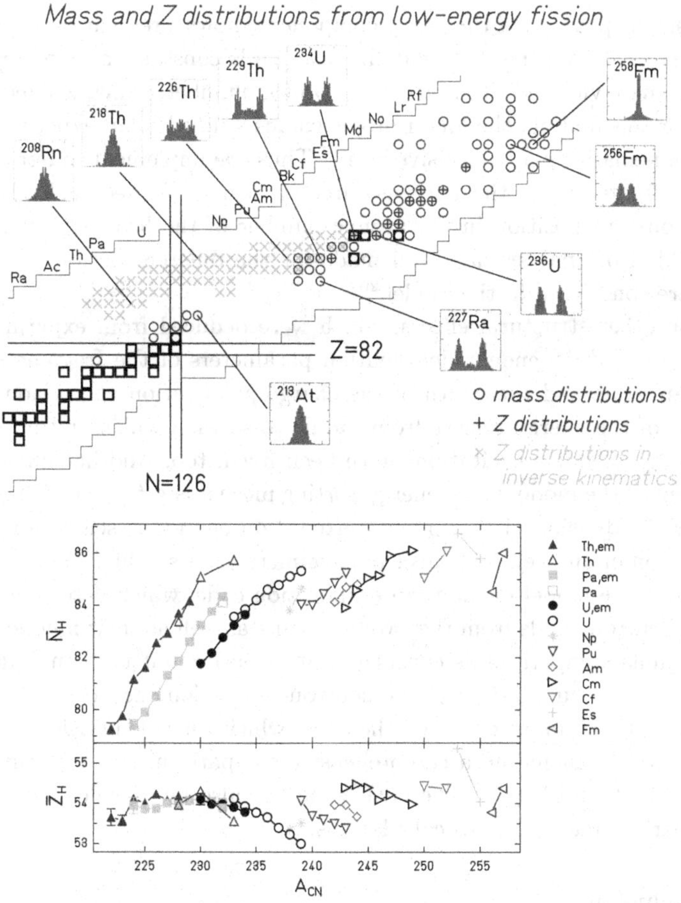

Mass and Z distributions from low-energy fission

Fig. 1. Upper part: Overview on the systems, for which mass or nuclear-charge distributions have been measured. The green crosses denote the systems which have been measured in inverse kinematics after electromagnetic excitation.[17] Lower part: The mean position of the asymmetric component in the heavy group in neutron number and in atomic number. Obviously, the traditional statement that the heavy component is constant at $A = 140$ must be revised by this analysis on a finer scale: The position of the asymmetric component is nearly constant at $Z = 54$, while the position in neutron or mass number varies by about 7 units.

heavy component of the fission-fragment distribution as an example of the many ingredients of the fission model. Fig. 1 shows the systems, for which mass or nuclear-charge distributions have been measured, on a chart of the nuclides. The position of the heavy component shows a regular pattern, but the previous assumption[18,19] that the position is constant in mass appears to be strongly violated. It is rather the proton number, which is fixed at $Z = 54$. For this finding, the long isotopic chains studied in an experiment in inverse kinematics play a decisive role.[17] Thus, we implement in our model that the freeze out of the mass-asymmetry degree of freedom leads to a nearly constant position in the atomic number of the heavy fragment. A similar kind of analysis has been made for the different fission channels, which are considered in the model.[20]

Some other structural effects, which were deduced from experimental data, are the Z-dependent deformation parameters of the fragments and the mean value and the width of the charge polarisation at scission. The fractions of the energy release from saddle to scission which end up in intrinsic and collective excitations have been fixed, too. Another important ingredient of the model is the energy-sorting mechanism,[21] which is responsible for the division of the intrinsic excitation energy at scission and for the creation of an even-odd effect in asymmetric mass splits.[22]

Finally, the model includes an evaporation code, which determines the prompt neutron yields from the two fragments as well as their kinetic energies. Gamma competition is considered; it smoothes out the consequences of the even-odd fluctuations of the neutron-separation energies.

More detailed information on the code, which we called GEF (GEneral Fission model), including a comprehensive comparison with experimental data can be found here.[23] The GEF code can also be downloaded, which allows performing dedicated calculations.

5. Conclusions

A new fission model has been developed. It is based on the statistical population of states in the fission valleys at the moment of dynamical freeze-out, which is specific to each collective degree of freedom. Three fission channels are considered. The separability principle governs the interplay of macroscopic and microscopic effects. The newly discovered energy-sorting mechanism determines the division of intrinsic excitation energy between the fragments at scission and the creation of a strong even-odd effect at large mass asymmetry. This new model gives a new insight into several dynamical times.

The GEF code provides a consistent description of the fission observables from polonium to fermium, from spontaneous fission to initial excitation energies up to about 14 MeV, with the same parameter set. (For higher excitation energies, multichance-fission must be considered.) Most parameters are fixed from independent sources, only less than 20 parameters have specifically been adjusted. Since the parameters of the model are closely related to physical properties of the systems, valuable conclusions on the fission process can be deduced. The good reproduction of measured data and the high predictive power of the code make it useful for applications in nuclear technology and complement the use of purely empirical models.

Acknowledgments

This work was supported by the EURATOM 6. Framework Programme "European Facilities for Nuclear Data Measurements" (EFNUDAT), contract number FP6-036434.

References

1. U. Brosa, S. Grossmann, and A. Mueller, *Phys. Rep.* **197**, 167 (1990).
2. A. C. Wahl, *Atom. Data Nucl. Data Tables* **39**, 1 (1988).
3. D. Lunney et al., *Rev. Mod. Phys.* **75**, 1021 (2003).
4. I. O. Morales et al., *Nucl. Phys. A* **828**, 113 (2009).
5. V. M. Strutinsky, *Nucl. Phys. A* **122**, 1 (1968).
6. M. Goeppert-Mayer, *Phys. Rev.* **75**, 1969 (1949).
7. D. Haxel et al., *Phys. Rev.* **75**, 1766 (1949).
8. W. D. Myers and W. J. Swiatecki, *Nucl. Phys. A* **612**, 249 (1997).
9. M. Goeppert-Mayer, *Phys. Rev.* **78**, 16 (1950).
10. D. Haxel et al., *Z. Phys.* **128**, 295 (1950).
11. C. Samanta and S. Adhikari, *Phys. Rev. C* **65**, 037301 (2002).
12. P. Moeller et al., *Phys. Rev. C* **79**, 064304 (2009).
13. B. D. Wilkins et al., *Phys. Rev. C* **14**, 1832 (1976).
14. K.-H. Schmidt et al., *Europh. Lett.* **83**, 32001 (2008).
15. U. Mosel and H. W. Schmitt, *Nucl. Phys. A* **165**, 73 (1971).
16. S. I. Mulgin et al., *Nucl. Phys. A* **640**, 375 (1998).
17. K.-H. Schmidt et al., *Nucl. Phys. A* **665**, 221 (2000).
18. J. P. Unik et al., *Proc. Symp. Phys. Chem. Fission, IAEA* **vol. 2**, 19 (1974).
19. Yu. Oganessian, *J. Phys. G: Nucl. Part. Phys.* **34**, R165 (2007).
20. C. Boeckstiegel et al., *Nucl. Phys. A* **802**, 12 (2008).
21. K.-H. Schmidt and B. Jurado, *Phys. Rev. Lett.* **104**, 212501 (2010).
22. B. Jurado and K.-H. Schmidt, these proceedings.
23. www.cenbg.in2p3.fr/GEF

THE STATISTICAL MODEL IN NUCLEAR FISSION-EXCITATION ENERGY AND SPIN POPULATION IN FRAGMENTS

HERBERT FAUST
Institut Laue-Langevin, 6 rue Jules Horowitz, 38042 Grenoble, France

We apply the statistical model of nuclear physics to the fission process, in particular to calculate excitation energy and spin distributions in fission products. We give the functions for these distributions, and by applying a Monte Carlo procedure we construct the distribution functions for the fragment kinetic energies. The temperature parameter which is needed for the distribution functions is calculated from an empirical law which relates nuclear temperature to the Q-value of the reaction. Results of the calculations are compared with experimental data, and excellent agreement is observed in all cases.

1. Introduction

Antagonistic views of fission are still persisting after more than 70 years of fission research. Geometric models try to map the fission path from the saddle point to scission by using microscopic-macroscopic models. Here shell correction values as function of deformation parameters play a fundamental role in establishing fission valleys in a multi-dimensional configuration space. Inclusion of a sufficient number of free parameters and constraints in quadrupole and octupole deformations allows subsequently establishing mass distributions. Fragment excitation within these models is mainly due to strong deformations along the fission path, and kinetic energy distributions are calculated via Coulomb repulsion of the deformed fragments. Angular momentum of the fragments is due to geometrical modes: bending, twisting, wiggling. In short, fragment excitation and angular momentum is calculated by evolution of the fragment shapes, and more precisely of the shapes at the very end of the fission process.

In contrast thermo-dynamical models try to apply statistical arguments to nuclear fission. Fission, as any other process in nature tries to maximize the entropy. The fissile actinide nucleus undergoes fission or any other nuclear decay because it is possible in this way to distribute energy which was concentrated in the actinide nucleus in many different ways. A multitude of fission products is created, and each fission product has many possibilities to share the available Q-value of the reaction between excitation and kinetic energy.

In the following we will apply the statistical model in a consequent way to nuclear fission, and see how far the model can predict fission variables.

Basic quantities to be calculated in nuclear fission are the mass and nuclear charge distribution of the primary fragments, their excitation energy and their spin. From the excitation energy distribution and spin distribution functions of the primary fragments further observables in the fission process can be calculated: the kinetic energy distributions by applying energy and momentum conservation laws, neutron and gamma emission from the well known statistical decay laws in nuclear physics, population of isomeric states and ground state band members in fission products, and the contribution of orbital momentum to the alignment.

2. The statistical model in nuclear fission

2.1. *The notion of temperature in nuclear physics*

Temperature in nuclear physics is defined by a distribution parameter kT which enters the following equation

$$P(E^*)dE^* = \frac{1}{Z}\exp(-E^*/kT) \cdot \rho(E^*)dE^* \qquad (1)$$

Here $P(E^*)$ is the probability that in a nuclear decay the decay product is excited to intrinsic excitation energy E^*. Important here are the Boltzmann factor which contains the temperature parameter, and the level density at this excitation energy $\rho(E^*)$. Z is the normalization of the integral of the distribution to 1.

Equation (1) is valid for a statistical ensemble of decay products of the same kind, and stemming from the same reaction. In nuclear fission the statistical ensemble comprises all fission fragments of one kind (same nuclear mass and nuclear charge) stemming from the same actinide nucleus which decays. This behavior is implicit in the level density expression which is dependent on (A,Z) of the decay product, and in the temperature which is dependent on the Q value of the reaction.

In contrast to the usual meaning of temperature we are dealing in nuclear fission with very short lived species, at an typical excitation energy of about 10 to 20 MeV nuclear lifetimes are of order 10(-17) seconds. This is the time a fission fragment takes part in the statistical ensemble which is looked at. Moreover the decays of the actinide nuclei happen uncorrelated in time and space, so that the decay products taking part in the ensemble are non-interacting, which is the characteristics of a micro-canonical ensemble.

2.2. Application of the statistical model in fission

It is well known in nuclear fission that the calculation of mass and charge distributions based on thermodynamic model poses problem. Due to the Q-value, which is mostly maximum for symmetric fragment split, mass distributions tend also to favor symmetry. Inclusion of fragment deformation at scission does not improve the agreement to experiment considerably. Our conclusion for charge and mass distributions would be that the statistical model here is not valid, unless the available phase space is truncated by barrier penetration effects and by selection rules. In the following we therefore address only to the energy and spin distribution of fission fragments. This approach implies that we can decouple the charge and mass distribution in the fission process from the distribution functions for excitation energy and spin, which leads to

$$W(A_i, Z_i, E_i^*, J_i) = \Theta(A_i, Z_i) \cdot \Phi(E_i^*, J_i) \qquad (2)$$

The probability of an actinide to divide into fragments $W(A_i, Z_i, E_i^*, J_i)$ with mass and charge (A,Z), excitation energy and spin (E_i^*, J_i) is separated in two functions, which are independent from each other. The subscript 'i' denotes which one of the two fission products is calculated. The function $\Phi(E_i^*, J_i)$ completely describes the distribution function for excitation energy and spin. It is this function which we will address to in the following. The statistical model allows a separation of this function into two parts, where the first one contains the excitation energy dependence and the second one the spin dependence of the fragment

$$\Phi(E_i^*, J_i) = P(E_i^*) \cdot G(J_i) \qquad (3)$$

This possibility of separating energy and spin stems from the structure of the level density expression $\rho(E^*, J)$ which is in general used in nuclear physics [1].

2.3. Distributions for excitation energy and kinetic energy of fragments

If the temperature of the statistical ensemble is known, Eq. (1) completely specifies the excitation energy distribution of the fragments. Level density expressions can be taken from the literature. In its simplest version the Fermi gas expression may be used

$$\rho(E_i^*) = \exp(2\sqrt{a_i E_i^*}) \qquad (4)$$

with a_i the level density parameter for fragment 'i'. Consequently the excitation energy distributions for both fragments read

$$P_1(E^*) \propto \exp(-E^*/kT) \cdot \rho_1(E^*) \qquad (5)$$

$$P_2(E^*) \propto \exp(-E^*/kT) \cdot \rho_2(E^*) \qquad (6)$$

for fragments 1 and 2, respectively. These distributions have a bell-shape character, with a maximum at $a_{1,2} \cdot (kT)^2$. The shape of both functions is different because in general the level density parameter is different for different fragments.

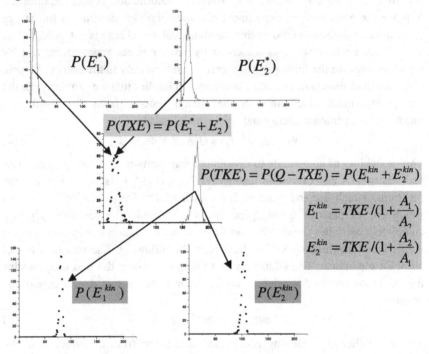

Figure 1. Monte-Carlo procedure to construct from the single fragment excitation energy distributions the TXE-distribution, and finally by applying momentum conservation the single fragment kinetic energy distributions. The example shows the calculation for 235U(n,f) splitting into 132Sn and 104Mo.

The above equations are for the distributions of the excitation energies. By applying energy and momentum conservation laws and using a sampling procedure it is possible to calculate from these expressions the distribution functions for the single fragment kinetic energies. The Monte-Carlo program samples randomly distribution from Eqs. (5) and (6), and constructs the distribution function for the total excitation energy

$$P(TXE) = P(E_1^* + E_2^*) \tag{7}$$

The Q-value for fission of the actinide (Ac, Zc) splitting into two fragments $(A_1, Z_1), (A_2, Z_2)$ is

$$Q = \Delta(A_c, Z_c) - [\Delta(A_1, Z_1) + \Delta(A_2, Z_2)] \tag{8}$$

The mass excesses Δ are taken from mass tables.
Applying energy conservation gives

$$P(TKE) = P(Q - TXE) \tag{9}$$

Fission is a binary process and we can apply momentum conservation which yields for the single kinetic energies of the fragments

$$E_1^{kin} = \frac{TKE}{1 + \frac{A_1}{A_2}}$$

$$E_2^{kin} = \frac{TKE}{1 + \frac{A_2}{A_1}} \tag{10}$$

The procedure is shown in Fig. 1.

2.4. Distribution functions for the fragment spins

For a sequence of shell model states the spin distribution $G(J_i)$ of a nucleus can be given as

$$G(J) = \exp(-J^2/2\sigma^2) - \exp(-(J+1)^2/2\sigma^2) \tag{11}$$

This function is approximated by a Gaussian with a weighting factor depending on the spin J

$$G(J) = \frac{2J+1}{2\sigma^2} \exp(-J(J+1)/2\sigma^2) \tag{12}$$

The spin cut-off parameter σ contains the mass of the nucleus, the level density parameter and the temperature of the statistical ensemble. Its value is given by Gilbert and Cameron [1] to be

$$\sigma^2 = 0.0888 \cdot a \cdot kT \cdot A^{2/3} \tag{13}$$

A typical spin distribution function for an A=100 fragment results in mean spin values of about 5 to 7 units of $\hbar$, and also very high spins of about $J = 15\hbar$ have a good chance of being excited.

2.5. An empirical relationship for the nuclear temperature value

Both distributions, the excitation energy and the spin distribution function, contain the nuclear temperature kT of the statistical ensemble. In general the temperature is derived from a decay law, which is mostly an expression which contains the kinetic energy of the emitted particles in a decay, the Q-value, and the level density. These dependencies are not yet known for fission, but it is possible to establish an empirical relationship between Q-value and temperature, see Faust and Bao [2].

$$kT = f \cdot Q \qquad (14)$$

We have determined the constant f for different actinide systems and we found a linear relationship

$$f = cZ_c + b \qquad (15)$$

For the actinide region in between Z=88 and Z=100 the values of the constants are $c = 3.29 \cdot 10^{-4}$ and $b = -0.0258$. With the value for the temperature of the ensemble given by Eqs. (14) and (15), respectively, all distributions for excitation energy and spin of the fragments are defined.

2.6. The choice of the level density parameter

Experimental level density parameter for fission fragments are available for spontaneous fission of 252Cf, measured by Butz-Joergensen and Knitter [3]. The measured values show the saw tooth behavior known from level densities from other reactions, and from theoretical approaches, where the level density parameter is calculated using a macroscopic-microscopic approach [1]. Level densities from nuclear fission do not appear to be different from other nuclear reactions, and we will use the parameters given by Gilbert and Cameron for our calculations. In order to take the dependence of the shell corrections on the excitation energy of the fragments into account we use the attenuation formalism given by Ignatyuk [4].

Level density parameter enters the equations for excitation energy and for the spin of the fragments. We use the Fermi-gas expression for the level density for all fragments. The appearance of the level density parameter in the expressions for the energy and spin distributions is the very reason why its saw tooth character with fragment mass is found again in mean fragment excitation, neutron multiplicity, and mean fragment spin.

3. Results and discussion

We have performed experiments on LOHENGRIN to measure single fragment kinetic energy distributions from thermal neutron induced fission of 233U. In Fig. 2 we show results of the measurements together with the calculations on the basis of Eqs. (5-10). The distribution functions agree well with each other for light and also for heavy fission products. In general the mean fragment kinetic energy is correctly calculated within few per cent, and also the higher moments of the distributions (variance and asymmetry) are well reproduced.

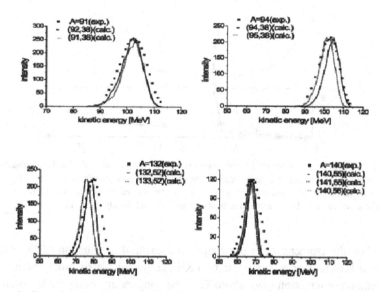

Figure 2. Examples for single fragment kinetic energy distributions. The points represent measurements from the LOHENGRIN spectrometer; the lines are the result of the calculations. Fragment masses A=91,94,132 and 140 are presented. The calculations have been done for different assumptions of nuclear mass and charge of the primary fragment which may be present in the measured lineshape.

In order to test validity of the spin expression of the statistical model we measured the population of isomeric levels in selected fragments in thermal neutron fission of 235U. The measurements were done in dependence of the single fragment kinetic energy. In the evaluation of the data assumptions have to be made on the neutron and gamma decay characteristics from the entry state given by Eqs. (5,6,12) to the respective isomeric levels. A typical decay pattern is shown in Fig. 3, where it is assumed that successive neutron emission and statistical gamma emission does not change the fragment spin. Only when the Yrast-line is reached discrete gamma emission takes away angular momentum. We used the expressions of England and Madland [5] to calculate the isomeric state population according to this picture.

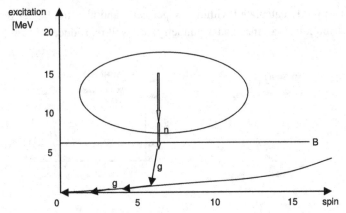

Figure 3. Decay pattern as assumed for a primary fission fragment. Entry states for the fragments are mostly within the oval borders as calculated by the distribution functions for excitation energy and spin. From each entry state neutron decay brings the fragment near to the neutron binding energy, from where on gamma decay dominates to the Yrast-line and finally along the Yrast-line to the ground-state.

When the isomeric state population is measured as function of the single fragment kinetic energy, the value of TXE can be calculated using energy and momentum conservation laws. From TXE the temperature value can be inferred, and in this case the calculation of mean fragment spin as function of the single fragment kinetic energy is parameter free, kT and level density parameter being known.

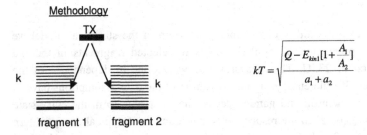

Figure 4. Methodology to calculate, from a single fragment kinetic energy measurement, the temperatue kT of the statistical ensemble. Note that also with a fixed value of the total excitation energy, the excitation of both fragments follow a temperature distribution.

It is worth mentioning that, also the value of TXE has a fixed value for a measured fragment kinetic energy, the excited states in both fragments still are distributed according to Boltzmann's law. Fig. 5 shows the result of the measurements of mean fragment spin as function of kinetic energy for 132Sb and

136I. The calculations are shown as solid line. They agree very well with the measured data points.

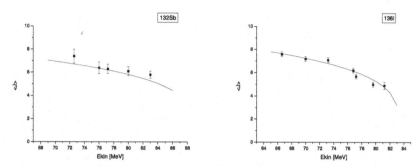

Figure 5. Examples for the population of isomeric states in nuclear fission. Shown is the mean value of the fragment spin as function of the single fragment kinetic energy for 132Sb and 136I. Data points are measurement from the LOHENGRIN spectrometer, the solid lines are the results from the calculation.

4. Conclusions

A separation of mass and charge distributions in binary fission from excitation energy and spin allows the application of the well known statistical model. The distributions, mean values and variances for excitation energy, kinetic energy and spins of fragments can be performed by employing an empirical relationship which allows for the determination of the temperature parameter. The nuclear temperature which characterizes the statistical ensemble of one kind of fission products, specified in mass and nuclear charge from a given fissioning system, is a linear function of the nuclear charge of the actinide system, at least for actinides in between Actinium and Californium.

References

1. A. Gilbert and A.G.W. Cameron, *Can. Journ. Phys.* **43**, 1446 (1965).
2. H. Faust and Z. Bao, *Nucl. Phys.* **A736**, 55 (2004).
3. C. Butz-Jorgensen and K.H. Knitter, *Nucl. Phys.* **A490**, 307 (1988).
4. A.V. Ygnatyuk, J.L. Weil, S. Raman and S. Kahane, *Phys. Rev.* **C47**, 1504 (1993).
5. D.G. Madland and T.R. England, *Nucl. Sci. Eng.* **64**, 859 (1977).

INVESTIGATION OF ^{234}U(n,f) AS A FUNCTION OF INCIDENT NEUTRON ENERGY

A. AL-ADILI*†, F.-J. HAMBSCH*, S. OBERSTEDT*, S. POMP†

EC-JRC - Institute for Reference Materials and Measurements, B-2440 Geel, Belgium
†*Division of Applied Nuclear Physics, UPPSALA UNIVERSITY*
751 20 Uppsala, Sweden

Measurements of the reaction ^{234}U(n,f) have been performed at incident neutron energies from 0.2 MeV to 5 MeV at the 7 MV Van De Graaf accelerator at IRMM. A twin Frisch-grid ionization chamber was used for fission-fragment detection. Parallel digital and analogue data acquisitions were applied in order to compare the two techniques. First results on the angular anisotropy and preliminary mass distributions are presented along with a first comparison between the two techniques.

Keywords: ^{234}U(n,f), Digital signal processing, Fission fragment mass distributions, Angular anisotropy.

1. Introduction

1.1. Motivation

In the reaction ^{234}U(n,f) the fissioning compound nucleus (CN) ^{235}U* corresponds to the second chance fission system in the reaction ^{235}U(n,f) and hence, information on the fission fragment (FF) characteristics are important for nuclear-reaction theory. This is, in particular, due to fluctuations in the FF properties observed in the sub-barrier region of the neutron-induced fission cross-section in the vicinity of the strong vibrational resonance at neutron energy around 800 keV [1]. A dedicated measurement campaign has been started to investigate those possible fluctuations in greater detail.

1.2. Digital and analogue processing

Beside the fission study of ^{234}U(n,f) a dedicated parallel data acquisition system was installed to compare conventional analogue acquisition (AA) and modern digital acquisition (DA) systems. Among the advantages of digital techniques is the drastic reduction in the amount of electronic modules and their replacement by software-wise implementation. This has the benefit of less electronic drifts during the experiment and the results are also less sensitive to calibration differences in gains and offsets. In addition, the post-experimental analysis' possibilities can be widely extended, because the whole digitized signal waveform is stored rather than only the pulse height (PH). Storage requirements are of course much more demanding in case of DA-systems, however nowadays no longer an issue.

2. Experiments

2.1. Measurements

Charged particles from the 7 MV Van De Graaf accelerator at the IRMM in Geel, Belgium were used to produce neutrons between 0.2 MeV and 5 MeV via the reactions ^{7}Li(p,n), ^{3}H(p,n) and D(d,n). A twin Frisch-grid ionization chamber (TFGIC) with P-10 as detection gas at a pressure of 1.05 bar is used to detect the FF [2]. The data acquisition is performed using a 100MHz, 12 Bit, wave-form digitizer for the digital system independently from the analogue electronic chains. The ^{234}UF$_4$ sample (92.133µg/cm^2) is vacuum deposited on gold-coated polyamide. For the absolute energy calibration the reaction ^{235}U(n_{th},f) is used for which the total kinetic energy (TKE) and the mean fragment mass number are well known and from which the angular distribution is isotropic. Three measurement series have been performed at neutron energies E_n = 0.2, 0.35, 0.5, 0.64, 0.77, 0.835, 0.9, 1.0, 1.5, 2.0, 2.5, 3.0, 4.0, 5.0 MeV. A precision pulse-generator was used to monitor and correct for electronic drifts in the AA system during the measurement time.

2.2. Ionization chamber

Both fragment energies and angles are measured in the TFGIC. Pre-neutron masses are then calculated iteratively using energy and momentum conservation. The TFGIC has two anodes, two grids and a common cathode, where the sample is located. The electrons induce a charge signal on the anodes which is emission-angle independent. The collected charge on the anode is proportional to the FF kinetic energy. The angle information is either calculated from the angle-dependent grid signal (summing method) or through the time difference between cathode- and anode-signal triggering (drift-time method). The summing method is sensitive to miss-calibrations due to the anode-grid summing. The drift time method is less straight forward to set up and has so far not been used in the analogue case. Therefore, in the following the summing method has been considered for both digital and analogue analysis.

2.3. Data acquisition

The electronic chain for one side of the TFGIC is shown in the upper part of Fig. 1. In the DA case only a timing-filter amplifier (TFA) and a constant-fraction discriminator (CFD) to generate the trigger from the cathode is needed. Otherwise the preamplifier signals are directly fed into the digitizer for pulse processing.

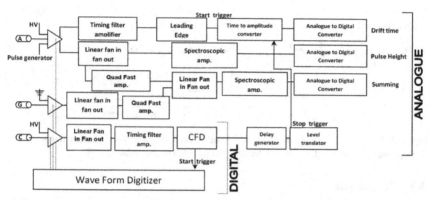

Figure 1. The electronics needed for one side of the chamber in the analogue and digital case.

3. Data analysis

3.1. Digital signal processing

In the digital case the pulse heights are determined by implementing the corresponding analogue modules as computer code in the digital off-line analysis program. The signals are first corrected for base-line shifts then differentiated in order to know the centre of gravity of the electron cloud. Also the preamplifier exponential decay ("ballistic deficit") is corrected for. The signals are then fed into a $CR-RC_4$ filter in order to reduce the pulse-to-noise ratio following the approach of Ref. [3]. The pulse heights are then created from the filter output. For the angle determination, the above procedure was applied to the sum of grid and anode signals, and then used as angle-dependent pulse height.

3.2. Pile-up correction

The main advantage of DA versus AA is the possibility of pile-up correction. The ^{234}U sample is highly α–active (150,000 α/sec) and one or two pile-ups are most probably coinciding with each fission event. A severe example is shown in

Fig. 2. In the digital case these wave-forms can be corrected, leaving a clean FF pulse without α-particle contribution to the pulse height. This correction also reduces the pile-up from neutron-proton elastic scattering due to the hydrogen in the P-10 counting gas.

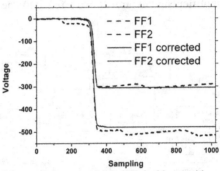

Figure 2. In this event three α-pile ups are visible in coincidence with one fission fragment. The pile-ups are corrected for digitally and the clean signal is left to determine the PH.

3.3. FF analysis

The PH signals are now undergoing the analysis described in Ref. [4]. Corrections are introduced for grid inefficiency, pulse height defect (PHD) and linear momentum transfer. The FF angle is calculated according to

$$\frac{X}{D}\cos\theta = \frac{A-\Sigma}{A} \qquad (1)$$

where X/D is the centre-of-gravity of the electron cloud distribution. A parabola is fitted to the half maximum of the $\cos(0°)$ and a linear fit to the $\cos(90°)$ distribution. The right hand side of Eq. (1) is divided by the range of the distribution between $\cos(0°)$ and $\cos(90°)$. The energy losses are assumed to be a linear function of $1/\cos(\theta)$. An extrapolation to channel zero leads to the crossing point between the two linear fits for both chamber sides, which corresponds to the case where the sample is infinitely thin. For ^{234}U(n,f) the neutron multiplicity as a function of mass, was calculated using the ratio of $\nu(^{233}U)$ and $\nu(^{235}U)$ from experimental data [5]. The TKE-dependency in $\overline{\nu}(A,TKE)$ was parameterized according to Ref. [6]. A correction for the neutron-energy dependency of total $\overline{\nu}$ was introduced based on experimental values from Ref. [7]. After all corrections the energy distributions from both chamber sides should coincide, which was verified. The FF masses are calculated by an iterative process. The thermal-neutron induced fission of ^{235}U was used for energy calibrate of ^{234}U(n,f). The recommended values for TKE (170.5 ± 0.5) MeV and mean heavy mass $<A_h>$ = 139.5 amu from Ref. [8] were used.

3.4. Angular anisotropy

The anisotropy of the angular distribution is studied in the centre-of-mass (CM) system relative to the isotropic distribution from the reaction ^{235}U(n_{th},f). The obtained ratio is then fitted for $0.3 \leq \cos(\theta) \leq 0.9$ with the two first even Legendre Polynomials P_0 and P_2 according to Eq. (2).

$$W(\theta) = A_0 \left(1 + A_2 \cdot \left(\frac{3 \cdot \cos\theta}{2} \right) \right) \quad (2)$$

The anisotropy, W(0°)/W(90°), is then given by Eq. (3).

$$\frac{W(0°)}{W(90°)} = 3 \cdot \frac{A_2}{2 - A_2} + 1 \quad (3)$$

4. Results

4.1. Angular distribution

The cosine distributions in the CM system both for AA and DA are in good agreement as shown for two measurements in Fig. 3. The angular resolution (FWHM) in terms of $\cos\theta_1 - \cos\theta_2$ obtained for ^{234}U at E_n = 2.5 MeV was 14.9% in the digital and 15.0% in the analogue case.

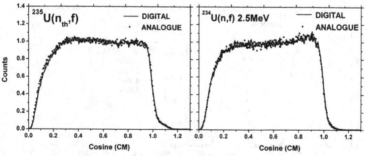

Figure 3. Angular distributions from digital and analogue analysis for two examples. In case of ^{234}U (2.5 MeV) one sees a slight positive anisotropy relative to the symmetry axis of the chamber.

4.2. Angular anisotropy

The angular anisotropy for the measured energies is presented in Fig. 4. The results are in good agreement with literature data showing a strong positive anisotropy of the fission fragment emission, around the vibrational resonance at 835 keV. At 500 keV the anisotropy turned strongly negative. Compared to previous studies the full angular range of nearly 2 x 2π is covered.

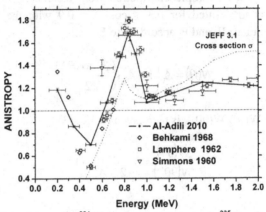

Figure 4. The angular anisotropy for ^{234}U(n,f) relative to the isotropic ^{235}U(n$_{th}$,f). Strong anisotropic behavior is observed around the vibrational resonance in agreement with literature data [9-11].

4.3. Mass distributions

The mass distribution from ^{235}U(n$_{th}$,f) is presented in Fig. 5a, both for DA and AA in comparison to literature data [12]. For ^{234}U(n,f) the first results show the expected increase of symmetric fission as a function of neutron energy (Fig. 5b).

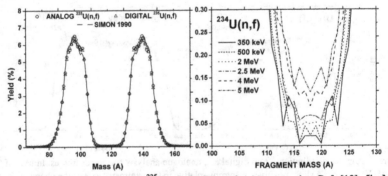

Figure 5a. The mass distribution for ^{235}U(n,f) for DA and AA compared to Ref. [12]. 5b. Mass distributions for ^{234}U(n,f) at different energies. Symmetric fission is increasing as a function of E$_n$.

5. Discussion

The present comparison between AA and DA shows, that the digital techniques are superior to the conventional ones. The results obtained so far are in both cases in very good agreement with literature data. Analogue data acquisition was far more sensitive to electronic drifts during the experiment than digital data acquisition. The possibility of individual pile-up correction is presently the main advantage with DA seen in this study, also in view of measuring possible other minor actinides. The pile-up correction gave a consistent explanation for the observed too high analogue pulse height after energy-loss correction relative to ^{235}U(n,f). In the digital data this was not experienced, because the α-contribution could be rejected. The strong angular anisotropy observed around the vibrational resonance indicates a change in fission properties at those incident neutron energies. Further FF analysis is ongoing.

Acknowledgements

We would like to thank the Van De Graaf staff for their support during beamtime. Special thanks go to I. Fabry and S. Zeynalov for providing guidance to the analogue data analysis and the digital signal processing, respectively.

References

1. A. A. Goverdovskii et al. *Soviet Journal of Nuclear physics*, **44**, 179–180, (1986).
2. O. Bunemann, et al, *Can. Jour. of res.* **A27**, 191, (1949).
3. O.V. Zeynalova, Sh.S. Zeynalov et al. У Д К 004.67 : 539.1.07, (2009).
4. C. Budtz-Jørgensen, et al. *Nucl. Instr. and Meth. in Phys. Res.* **A258** 209-220, (1987).
5. C. Wahl, *Atomic Data and Nuclear Data Tables* **39**, 1-1 56 (1988).
6. G. Barreau, *Nucl. Phys.* **A432**, 411-420, (1985).
7. D.S. Mather et al. *J,NP,66,149,6504* (1965).
8. C. Wagemans. In *The Nuclear Fission Process*, 323, (1991).
9. J. E. Simmons, et al, *Phys. Rev.* **120**, 198 (1960).
10. R. Lamphere, *Nucl. Phys.* **38**, 561-589, (1962).
11. A. N. Behkami et al, *Phys. Rev.* **171**, Number 4, 170, (1968).
12. G. Simon, et al, *Nucl. Instr. and Meth. in Phys. Res.* **A286**, 220-229, (1990).

MEASUREMENT OF FRAGMENT MASS YIELDS IN NEUTRON-INDUCED FISSION OF ^{232}TH AND ^{238}U AT 33, 45 AND 60 MEV

V.D. SIMUTKIN*, S. POMP, J. BLOMGREN, M. ÖSTERLUND, P. ANDERSSON
and R. BEVILACQUA

*Department of Physics and Astronomy, Uppsala University,
Uppsala, SE-75120, Sweden
* E-mail: vasily.simutkin@fysast.uu.se
www.fysast.uu.se*

I.V. RYZHOV and G.A. TUTIN

*V.G. KHLOPIN Radium Institute,
2nd Murinskiy prospect 28, Saint-Petersburg, RU-194021, Russia*

M.S. ONEGIN and L.A. VAISHNENE

*Petersburg Nuclear Physics Institute,
Gatchina, Leningrad district, RU-188300, Russia*

J.P. MEULDERS and R. PRIEELS

*FNRS and Institute of Nuclear Physics,
Université catholique de Louvain, B-1348 Louvain-la-Neuve, Belgium*

Over the past years, a significant effort has been devoted to measurements of neutron-induced fission cross-sections at intermediate energies but there is a lack of experimental data on fission yields. Here we describe recent measurements of pre-neutron emission fragment mass distributions from intermediate energy neutron-induced fission of ^{232}Th and ^{238}U. The measurements have been done at the quasi-monoenergetic neutron beam of the Louvain-la-Neuve cyclotron facility CYCLONE and neutron peak energies at 32.8, 45.3 and 59.9 MeV. A multi-section Frisch-gridded ionization chamber was used as a fission fragment detector. The measurement results are compared with available experimental data. Some TALYS code modifications done to describe the experimental results are discussed.

Keywords: ADS, neutron-induced fission, fission fragment mass yield, intermediate energy

1. Introduction

The operation of Accelerator Driven Systems (ADS) is based on the neutron-induced fission of actinides at neutron energies up to 1 GeV. It implies that the nuclear data in this energy range are extremely important for the use of ADS. In particular, neutron-induced fission yields are important for two reasons. The knowledge of a fission product composition of a spent ADS fuel is necessary in taking decisions about possible actions with the spent fuel. Second, fission products, by absorbing and emitting neutrons, influence the neutron balance of a system. Therefore, fission product yields have to be considered already at the design stage of an ADS. However, there is a lack of data on fission yields at intermediate energies. For instance, ^{238}U and ^{232}Th are two fertile isotopes which should be taken into consideration in possible designs of ADS. But, to date, only one measurement of intermediate energy neutron-induced fission fragment mass yields of ^{238}U has been done[1] and no experimental data exist for ^{232}Th. In our experiments, we have measured the pre-neutron emission fragment mass distributions in neutron-induced fission of ^{232}Th and ^{238}U for energies 32.8, 45.3 and 59.9 MeV. The thorium data have been obtained for the first time.

2. Experiment

The experiments have been done at the quasi-monoenergetic neutron beam of the Louvain-la-Neuve cyclotron facility CYCLONE.[2] The neutrons were produced by a monoenergetic proton beam impinging a 5 mm thick Lithium target. A multi-section Frisch-gridded ionization chamber (MFGIC) similar to that described in[3] was used for the detection of fission fragments. The fissile targets were prepared by vacuum evaporation of ^{nat}UF$_4$ and ^{232}ThF$_4$ onto 30 μg/cm^2 thick formvar backings. The backings were covered by 10-15 μg/cm^2 Au to make them electrically conducting. The average thickness of the fissile targets was 130 and 70 μg/cm^2, respectively, for the thorium and uranium deposits. The first fissile target was located at a distance of 375 cm from the Li target. At the target positions, the fluence rate of peak neutrons was about 10^5 cm^{-2}s^{-1}.

3. Data processing and analysis

Having the energies of the complementary fragments in the c.m. system one can determine the fragment masses using the double kinetic energy method based on the conservation laws of mass and linear momentum. Ideally, the pre-neutron emission fragment masses m_i^* (i=1,2) are calculated as

$$m_1^* = A_{cn} \frac{E_2^*}{E_1^* + E_2^*}, \quad m_2^* = A_{cn} - m_1^* \ , \tag{1}$$

with A_{cn} the mass number of the compound nucleus and E_i^* the pre-neutron emission fragment kinetic energies in the c.m. system. In practice, however, only post-neutron emission fragment kinetic energies E_i can be measured. By assuming isotropic neutron emission from fully accelerated fragment the pre-neutron kinetic energy is given as

$$E_i^* = E_i \left(1 + \frac{\nu_i(m_i^*)}{m_i} \right) \ , \tag{2}$$

where $\nu_i(m_i^*)$ is the average number of neutrons emitted from the fragment and m_i is the post-fission fragment mass:

$$m_i = m_i^* - \nu_i(m_i^*) \ . \tag{3}$$

From Eqs. 1-3 the pre-neutron fragment masses and energies were determined in an iterative procedure. The post-neutron fragment energy in the lab-system was obtained from the anode pulse height P_{anode} as:

$$E_i^{lab} = A_\alpha P_{anode} + \Delta_{grid} + \Delta_{PHD} + \Delta_{loss} \ , \tag{4}$$

where A_α is the constant obtained from an absolute energy calibration using α-particles from the ^{252}Cf source and a high precision pulse generator, Δ_{grid}, Δ_{PHD} and Δ_{loss} are the corrections for the grid inefficiency, the detector pulse-height-defect (PHD) and the fragment energy losses in the target material and backing, respectively. The post-neutron fragment energy in the c.m. system was calculated as

$$E_i = E_i^{lab} + \frac{p_F^2}{2A_F} - 2p_F \left(\frac{E_i^{lab}}{2A_F} \right)^{\frac{1}{2}} \cos\theta \ , \tag{5}$$

where p_F is the average longitudinal linear momentum of the fissioning nucleus, ν_{pre} is the pre-fission neutron multiplicity and θ is the fragment emission angle with respect to the normal of the cathode plane. The time-of-flight (TOF) technique was applied to identify the fission events induced by peak neutrons. To substract the wrap-around background, the fission event time distributions were simulated by a Monte-Carlo folding the neutron-induced fission cross-sections of ^{232}Th[4] and ^{238}U[5] with the neutron spectrum[2] (see Fig. 1a, left part).

The evaluation[6] was then used to calculate the mass distributions corresponding to the overlapping low energy neutrons.

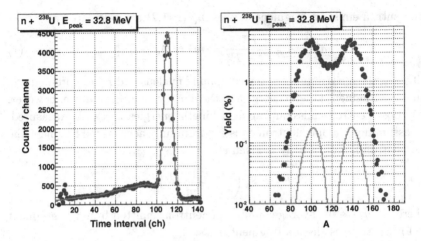

Fig. 1. (a) Left part: measured (blue symbols) and simulated (red line) TOF spectra of ^{238}U fission events for $E_n = 32.8$ MeV. (b) Right part: wrap-around contribution (continuous line) into the pre-neutron emission fragment mass-distribution.

The mass resolution function was calculated as in:[7]

$$\sigma_{TOT}^2(m) = \sigma_{INH}^2 + \sigma_{INST}^2 ,\qquad(6)$$

where σ_{INH}^2 is the inherent broadening due to prompt neutron emission and σ_{INST}^2 is the instrumental broadening caused by the measurement technique. The inherent mass dispersion can be written as

$$\sigma_{INH}^2(m) = \sigma_{FF}^2 + \sigma_{CN}^2 + \sigma_{LMT}^2 ,\qquad(7)$$

where σ_{FF}^2 defines the mass variation due to neutron emission from the fission fragments[1], σ_{CN}^2 is the mass variation due to the pre-fission neutrons[1] and σ_{LMT}^2 approximates the variance in mass caused by dispersion of the linear momentum transfer distribution.[8] The total fragment mass resolution was found to be in the interval 8-10 amu depending on the neutron energy. Using the estimated mass resolution, the final mass distributions were found in an iteration procedure.

4. Measurement results

Pre-neutron emission fragment mass yields measured for the ^{232}Th(n,f) and ^{238}U(n,f) reactions at 32.8, 45.3 and 59.9 MeV are presented in Fig. 2 and Fig. 3.

Figure 3 shows that our ^{238}U results at 32.8 and 45.3 MeV agree well with the Zoller data[1] while it is worse at 59.9 MeV. For a proper comparison,

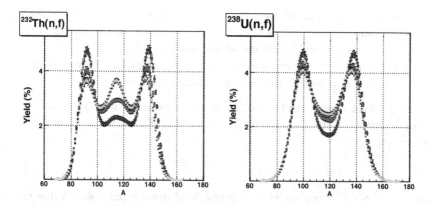

Fig. 2. Pre-neutron emission fragment mass distributions measured for neutron-induced fission of ^{232}Th (left) and ^{238}U (right) at 32.8, 45.3 and 59.9 MeV. The yield of the symmetric component increases with the raise of the neutron energy.

it should be realized that we detected fragments within a cone around the neutron beam axis, while the other measurement was done at fragment emission angles close to 90 degrees.

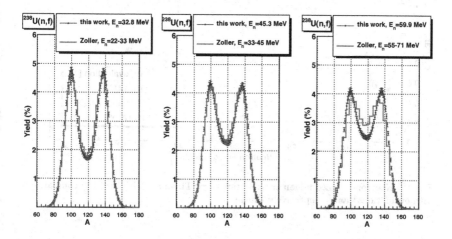

Fig. 3. The present data for ^{238}U in comparison with the existing data.[1]

The experimental mass yields will also be compared with TALYS[9] calculations. As a first step, we described the neutron-induced fission cross-sections of ^{238}U and ^{232}Th . Several modifications of the TALYS 1.0 code have been done to get better agrement with the evaluated data:

- Nuclear level densities of the deformed states have been considered separately and independently from the ground state level density for each Th and U isotope which undergoes fission.
- We disabled the use of collective enhancement factors for the calculation of level densities of the deformed states.

To get better agreement with the experimental data, we varied parameters $\tilde{a}$ and γ of the formula by Ignatyuk et al.[10] for the level density parameter a

$$a = a(E_x) = \tilde{a}\left(1 + \delta W \frac{1 - exp(-\gamma U)}{U}\right) , \qquad (8)$$

where $\tilde{a} = a(E_x \to \infty)$ is the asymptotic level density parameter value in the absence of shell effects and γ is the damping parameter which describes how rapidly $a(E_x)$ approaches a. Our results of the neutron-induced fission cross-sections versus neutron energy are shown in Fig. 4.

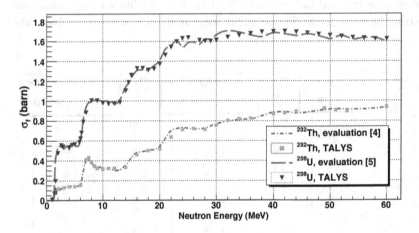

Fig. 4. Neutron-induced fission cross-section of ^{232}Th and ^{238}U (points) in comparison with evaluations[4] and[5] respectively (lines)

5. Conclusions

We have measured neutron-induced fission fragment mass yields of ^{238}U and ^{232}Th at the incident neutron energies 32.8, 45.3 and 59.9 MeV. The thorium data have been measured for the first time. The experimental data have been presented at the ND2010 conference.[11]

Acknowledgements

The authors thank the operating crew of the Louvain-la-Neuve cyclotron facility for the excellent neutron beams they provided for these experiments. This work was supported in part by the International Science and Technology Center (project 3192) and the European Commission within the Sixth Framework Programme through I3-EURONS (contract no. RII3-CT-2004-506065).

References

1. C. Zöller, Untersuchung der neutroneninduzierten spaltung von ^{238}U im energiebereich von 1 MeV bis 500 MeV, PhD thesis, TH Darmstadt, (Darmstadt, 1995).
2. H. Schuhmacher, C. Brede, V. Dangendorf, M. Kuhfuss, J. Meulders, W. Newhauser and R. Nolte, *Nucl. Instrum. Meth.* A **421**, 284 (1999).
3. I. Ryzhov, G. Tutin, A. Mitryukhin, V. Oplavin, S. Soloviev, J. Blomgren, P. Renberg, J. Meulders, Y. E. Masri, T. Keutgen, R. Prieels and R. Nolte, *Nucl. Instrum. Meth.* A **562**, 429 (2006).
4. M. Chadwick, P. Obložinský, M. Herman *et al.*, *Nuclear Data Sheets* **107**, 2931 (2006).
5. V. Pronyaev, S. Badikov, A. Carlson *et al.*, in *Proc. of IAEA Co-ordinated Research Project on International Evaluation of Neutron Cross-Section Standards*, (STI/PUB/1291) (International Atomic Energy Agency, Viena, 2007).
6. D. Gorodisskiy, K. Kovalchuk, S. Mulgin, A. Rusanov and S. Zhdanov, *Ann. Nucl. Energ.* **35**, 238 (2008).
7. L. Glendenin, J. Unik and H. Griffin, Determination of primary nuclear charge of fission fragments from their characteristic k-x-ray emission in spontaneous fission of ^{252}cf, in *1st IAEA Symposium on Physics and Chemistry of Fission*, (Vienna, 1965).
8. M. Fatyga, K. Kwiatkowski, H. Karwowski, L. Woo and V. Viola, *Phys. Rev.* C **32**, 1496 (1985).
9. A. Koning, S. Hilaire and M. Duijvestijn, Talys: Comprehensive nuclear reaction modeling, in *Proc. Int. Conf. on Nuclear Data for Science and Technology*, (AIP Conference Proceedings 769, Santa Fe, USA, 2005).
10. A. Ignatyuk, G. Smirenkin and A. Tishin, *Sov. J. Nucl. Phys.* **21**, p. 255 (1975).
11. I. Ryzhov, V. Simutkin, G. Tutin, M. Onegin, L. Vaishnene, J. Blomgren, S. Pomp, M. Österlund, P. Andersson, R. Bevilacqua, J. Meulders and R. Prieels, Measurement of fragment mass yields in neutron-induced fission of ^{232}Th and ^{238}U at 33, 45 and 60 MeV, in *Proc. Int. Conf. on Nuclear Data for Science and Technology*, (Jeju Island, Korea, 2010).

PROGRESS IN THE ATOMIC NUMBER IDENTIFICATION OF FISSION FRAGMENTS

I. TSEKHANOVICH*, A.G. SMITH, J.A. DARE, A.J. POLLITT

Department of Physics and Astronomy, University of Manchester,
Oxford road, Manchester M13 9PL, UK
* E-mail: igor.tsekhanovich@manchester.ac.uk

This paper discusses the problem of nuclear-charge identification of non-accelerated fission products by non-radiative methods. The major factors influencing Z resolution are pointed out, as well as the ways to overcome them. Two basic techniques for nuclear charge assignment are presented and discussed from the viewpoint of fast digital sampling electronics used in conjunction with specially developed algorithms at the Manchester 2v2E spectrometer: (1) analysis of the specific energy losses of fragments and (2) correlation of their mass, energy and range with atomic number. Fragments kinetic energy and range parameters are obtained, and specific energy loss is deduced, from the analysis of the fragments pulse shapes and lengths in gaseous detectors. Special attention is given to the range measurement and analysis; preliminary results on the deduced average nuclear charge parameter are demonstrated and discussed.

Keywords: Fission Fragment Identification; Specific Energy Loss; Mass-Energy-Range Correlation; Pulse Trace Analysis.

1. Introduction

Identification of atomic numbers of non-accelerated fission products by physical, non-radiative, methods is a long-standing experimental problem. It is well-known that, in low-energy fission, fission products do not possess enough kinetic energy to produce a Bragg peak when travelling in a stopping medium (i.e., in a detector). This rules out the so-called Bragg curve spectroscopy[1] as a method for the nuclear charge assignment of fission fragments. Despite the absence of the Bragg peak, the amount of ionisation created by nuclei per unit of length of their track in the detector still carries information on the atomic number. The main difficulty in the analysis of this (specific) ionisation is that the difference in the amount of energy lost along certain fixed length of the track by fragments with adjacent nuclear charges is small and comparable to the detector energy resolution.

An ionization chamber (IC) is best suited detector to measure kinetic energy of fission products. This is due to its 100% efficiency for nuclei and insensitivity to nuclear irradiation. In addition, the use of an IC as a fission-fragment detector allows us to know and control the thickness of the dead layer (IC entrance window), which is not the case for the other types of detectors. The actual (measured) resolution on energy of an IC can be considered as originating from the three sources: (i) the intrinsic resolution coming from the statistical nature of the fragment-gas interaction (ionising and non-ionising parts) and the collection of the produced ionization; (ii) the fragments energy straggling in the insensitive part of the detector (entrance window) and (iii) signal loses and noise in the processing electronics chain. Therefore, reliable assignment of atomic numbers to fission products may only be possible if thorough care is given to the construction of the IC and to the collection of the ionization produced in it, as well as to the consequent electronic treatment of signals. This approach is implemented in the design of the two-arm spectrometer of fission products STEFF[2] built at the University of Manchester. In addition to the identification of masses, the spectrometer will deliver information on nuclear charges of fission products. This will be achieved from the analysis of the fragments pulse shapes and ranges in gaseous detectors, both obtained using digital electronics in conjunction with specially developed algorithms.

2. Nuclear charge identification methods

Two methods can be used to assess the atomic numbers of fragments created in low-energy fission reactions:

- Analysis of specific energy losses, and
- Correlation between fission-fragment mass, energy, range and atomic number.

2.1. *Analysis of specific energy losses*

This method known also as $\Delta E - E$ technique has its origin in the LSS theory[3] and been successfully used by different research groups for nuclear charge assignment of low- and high-energy fission products for more than four decades. Considering the stopping medium parameters as constant, LSS makes the following link between the specific energy loss, dE/dx, and parameters of stopping nuclei such as velocity v and atomic number Z:

$$\frac{dE}{dx} = f(v, Z) \qquad (1)$$

Eq.1 gives the methodology to obtain Z, which is to inspect specific energy losses for a fixed velocity value and for nuclei from the same isobaric chain. Obviously, the spread in kinetic energies of the selected isobars due to the velocity bin width should be held as low as possible in order to achieve the best conditions for the Z identification. This assumes availability of a large set of data so that the narrow gating on velocity provides sufficient statistical accuracy for this kind of analysis. In the case of the Manchester spectrometer operating at present with a ^{252}Cf fission-fragment source, the necessary accuracy of the measurement can be achieved only in about a week of continuous running; this is due to the source strength and the solid angle of the device.

It should be stressed that the outlined technique appears to have a resolving power limit on the level of $Z/\Delta Z \sim 40 - 42$ (see, for instance, Ref. 4). It is however not clear if there exists a principal limiting factor making resolution of nuclear charges impossible above a certain Z value. One can therefore hope to improve on the Z identification by pushing energy and time resolution of detectors to their limits, and taking thorough care about electronic noise suppression and electronic pulse processing. Whereas the former was possible already for some time - here it is pertinent to cite the Cosi-fan-Tutte spectrometer[5] as example for the device with an unprecedented precision on energy and time measurement — the latter became possible only recently, with the development of digital electronics. The advantage of digital techniques for the Z identification purposes is that the full shape of ionization traces of fission products can be sampled and stored for further off-line dE/dx analysis. The importance of digital sampling of dE/dx curves has been realized by F. Gönnenwein already in mid nineties when his team experimented at the ILL mass separator with digitizing oscilloscopes and flash ADCs.[6] However the exclusion of pulse-processing analogue modules from the electronic circuit assumes that the off-line treatment has to be given to the stored pulse traces in order to be able to perform any comparative dE/dx analysis. The basics of digital pulse treatment algorithms developed over the last four years at Manchester are described in the present Proceedings.[7]

With carefully chosen IC geometry (thin (0.5μm) entrance window, size of the chamber, size of the anode segments, minimized parasitic capacitances, optimized gas pressure), suppression of induced RF noise (preamplifiers in the IC vessel, screening of signal cables, optimized noise filters parameters), and the appropriate pulse treatment[7] the Manchester 2v2E spectrometer is expected to be able to demonstrate some progress in the nuclear charge identification. The limiting factor for today is the stability of the operation of the TOF STOP detectors. They are gas-filled MWPCs which appear to be very sensitive to the day-night temperature changes. Even small drifts in pressure due to temperature changes affect the timing resolution of the TOF section of the spectrometer thus making precise velocity gating very difficult. Therefore the structures seen initially in the dE/dx spectrum for a selected mass and velocity get smeared out with time. Efforts are currently being undertaken for both trying to correct for the timing drifts of the STOP detector and for improving its temperature stability.

2.2. Mass-energy-range correlation

This method is based on measurements of fission-fragment kinetic energies E and ranges R in the detector. From the N. Bohrs theoretical approach,[8] the following approximate formula can be used to deduce atomic number from the measured parameters:

$$R \propto \frac{\sqrt{A \cdot E}}{Z^{2/3}} \qquad (2)$$

As mass is not the observable it can be replaced with the time of flight between the spectrometer START and STOP detectors and kinetic energy. Rearrangement for Z gives:

$$Z^{2/3} \propto \frac{ToF \cdot E}{R} \qquad (3)$$

According to Eq. 3, the three observables have equal importance for the Z determination of fission products; and so will be equally important for the Z resolving power. It should however be noted that the precision on the range parameter ΔR is dependent on that of the kinetic energy as range R is deduced from the sampled pulse traces. We deduce fission-fragment ranges as difference between two thresholds on the total ionization trace. The upper threshold can be set as high as 100% of the energy value, whereas attention should be paid to the lower threshold to be distinctly above the baseline noise level. Precision to which upper and lower thresholds are determined is given by two factors: (i) sampling rate of the electronics and (ii) quality of the pulse trace: it should be free of noise ripples.

Kinetic energy and range are the parameters which do not have a memory as to what happened to fission fragments on their way to the energy detector; one may therefore expect a strict linear correlation between them.

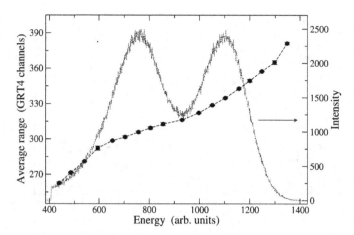

Fig. 1. Filled symbols: Average range of fission fragments as function of their kinetic energy. Dots depict the kinetic energy distribution as measured in the ionisation chamber.

Unfortunately not all the kinetic energy of fragments is converted into ionization in the detector body. There are non-ionizing interactions known as Pulse-Height Defect (PHD) whose strength is dependent on the nature of slowing down fragments.

Fig. 1 demonstrates energy dependence of average range values deduced from small energy bins on the measured kinetic-energy distribution (only full-energy hits in the central anode of the IC are considered). Generally, there is a change in the slope in the range/energy correlation for the two fragment groups. The effect is thought to be due to the PHD. The sharp decrease in average range in the low-energy wing is artificial; it comes from the condition set on the lower threshold, truncating the real range spectra. The lightest fission fragments corresponding to the highest kinetic energy values seem to have an enhanced range value than expected from the trend in the light-fragment group. This has yet to be understood; the effect may have links to the structure of nuclei in the region and therefore requires more careful experimental investigation involving range analysis as a function of mass for different source/target nuclei.

Dependence of range on the fission-fragment mass is given in Fig. 2. Due to the imperfect TOF detectors the mass resolution of the spectrometer is estimated to be at about 2-3 mass units, therefore the mass values in Fig. 2 should be considered as having only an indicative character.

In accordance with Eq.2, one would expect the range parameter to be quasi constant in the case of light fragments, as for this fragment group E

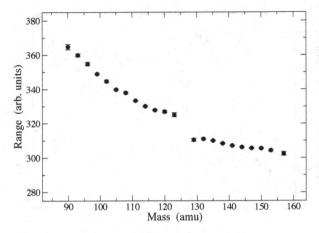

Fig. 2. Average range of fission fragments vs. fission-fragment mass. Considered only full energy hits in the central anode of the ionisation chamber. The mass values are indicative ones.

practically does not vary with mass and some increase in A is compensated by the increase in Z; for the heavy fragment group the behavior of range would be dominated by the fall in the kinetic-energy parameter.

Surprisingly, the measured results show a behavior in contradiction to the expected one (see Fig.2). Inclusion of the PHD correction into the energy (=mass) determination is very unlikely to explain the downsloping trend and the freeze-out in ranges for the light- and heavy-fragment groups, correspondingly. Consequently, it should be assumed that Eq.2 does not provide a correct relationship between the four parameters. Investigations are under way to understand the range parameter behavior at different fixed TOF and/or energy values in order to finally obtain a better match with the experimental observation.

Despite the disagreement between Eq.2 and experimental data on ranges, it is nevertheless worthwhile looking at how the atomic number parameter will behave if calculated from Eq.3. The obtained results are presented in Fig.3 as a function of mass number. Astonishingly good linear dependence of average Z on the mass number immediately catches the eye. The linear trend appears to persist for a big fraction of the light-mass peak, and starts to fade away only for fragments with masses $A > 110$; the remnants of this sensitivity to atomic number are still visible is the vicinity of $A = 132$. Nevertheless one should state that, quantitatively, Eq.3 fails to identify average atomic number for heavy fragments as it gives values

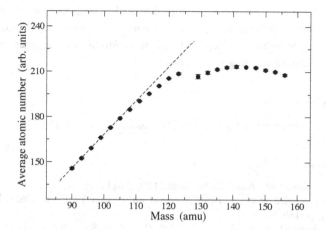

Fig. 3. Average atomic number vs. fission-fragment mass. The dashed line is just to guide the eye.

lower than those for the heaviest light nuclei. On the other hand, it has to be stressed that no gating has been applied to obtain results plotted in Fig. 3; the whole bulk of the data has been made use of. The observed sensitivity to the atomic number makes the method very interesting for identification of fission products in general and for the γ-ray spectroscopy studies in particular.

A fragment-γ-γ coincident measurement is currently going at the Manchester laboratory. It is hoped to be able to make a selection of nuclei by their atomic number and mass using known prompt γ-ray transitions. This would provide clean conditions for investigation of R/A and R/Z dependencies, as well as a possibility to test the technique for the nuclear-charge resolving power.

3. Conclusion

We have briefly described two approaches used at the Manchester fission-fragment spectrometer for identification of atomic numbers of non-accelerated fission products. They are based on the analysis of specific energy losses and mass-energy-range correlation of fission products. Though both techniques have been known for many years, only the former could be widely experimentally exploited, mainly due to the lack of methods for precise range measurements. An attempt has been made to improve on the nuclear charge assignment through the development of the latter technique. The results presented in this paper are quite preliminary as not all of the spectrometer detectors yet function with a desired level of precision. However they give room to believe that certain progress is still possible to make in the domain. The expectations lie in further development and use of digital pulse-processing techniques.

Acknowledgment

This work is supported by the EPSRC research grant 2-4570.5, UK.

References

1. C.R. Gruhn et al., Nucl. Instr. Meth. **196**, 33 (1982).
2. I. Tsekhanovich et al., 1. Proc. Seminar on Fission VI, Corsendock, Belgium (2007), p. 189.
3. J. Lindhard et al., 1. Mat. Phys. Med. Dan. Vid. Selsk. **33**, No.14 (1963).
4. J.P. Bocquet et al., Nucl. Instr. Meth. **A267**, 466 (1988).
5. N. Boucheneb et al., Nucl. Phys. **A502**, 261c (1989).
6. F. Gönnenwein et al., ILL experimental reports *3-01-293* (1995), *3-01-293* (1996) and *3-01-293* (1997.)
7. A.J. Pollitt, in This Proceedings Book.
8. N. Bohr, Phys. Rev. **59**, 270 (1941).

FRAGMENT CHARACTERISTICS FROM PHOTOFISSION OF ^{234}U AND ^{238}U INDUCED BY 6.0 - 9.0 MEV BREMSSTRAHLUNG

A. GÖÖK*, R. BARDAY, M. CHERNYKH, C. ECKARDT, J. ENDERS,
P. VON NEUMANN-COSEL, Y. POLTORATSKA, M. WAGNER

Institut für Kernphysik, Technische Universität Darmstadt,
D-64289, Germany
*E-mail: agook@ikp.tu-darmstadt.de

A. RICHTER

Institut für Kernphysik, Technische Universität Darmstadt and ECT*,
Villazano(TN), I-38123, Italy

S. OBERSTEDT, F.-J. HAMBSCH

EC-JRC, Institute for Reference Materials and Measurements (IRMM)
Geel, B-2440, Belgium

A. OBERSTEDT

School of Science and Technology, Örebro University
S-70182, Sweden

As a preparatory experiment for a search for parity violation in photofission, fission of ^{238}U and ^{234}U induced by 6 – 9 MeV bremsstrahlung has been investigated at the superconducting Darmstadt electron linear accelerator S-DALINAC. Using a twin Frisch grid ionization chamber fission fragment energy and mass distributions have been determined by means of the double kinetic energy technique. The experiment was performed in order to test the ionization chamber's performance in a bremsstrahlung environment. Results on the fission fragment characteristics from the ^{238}U(γ,f) reaction are found to be in good agreement with literature values. In addition results on fission fragment mass and energy distributions from the ^{234}U(γ,f) reaction are presented for the first time in this energy region.

Keywords: Parity violation; Photofission; Ionization chamber; ^{238}U(γ,f); ^{234}U(γ,f).

1. Introduction

Due to the weak interaction parity-non-conservation (PNC) can be observed in nuclear reactions as spatial reflection asymmetries. The magnitude of the asymmetry can be related to the ratio of the weak and strong coupling constants, and such an estimate results in an order of magnitude of $\sim 10^{-7}$. However, asymmetries several orders of magnitude higher have been measured in compound nucleus reactions. Very large asymmetries in the order of a percent have been detected in the helicity-dependent transmission of polarized neutrons from a variety of targets ranging from ^{81}Br to ^{238}U [1]. The experimental findings indicate that kinematic and dynamic enhancement factors play an important role [2]. In order to resolve open questions [1, 3] in describing the enhancement factors, further experiments with different probes are helpful.

The new polarized injector SPIN [4], currently under installation at the superconducting Darmstadt electron accelerator S-DALINAC [5], will be able to deliver a high intensity beam of longitudinally polarized electrons. This allows investigating PNC in fission induced by circularly polarized bremsstrahlung, created from the longitudinally polarized electron beam.

Since in binary fission the two fragments fly apart in opposite directions, the PNC asymmetry has to be measured for specific fragments. Hence, in order to detect PNC, the fission detector must be able to simultaneously determine the masses of the fission fragments and the orientation of the fission axis. A further requirement on the detector is that it is able to withstand the high integrated luminosity needed to collect the necessary statistics. A good approach seems to be the use of a twin Frisch grid ionization chamber, which is radiation hard when operated under constant gas-flow. The experiment presented in the following was undertaken in order to study the detector's performance in the bremsstrahlung setup at the S-DALINAC injector.

2. Experiment

The experiment was performed at the S-DALINAC injector, which can deliver an electron beam with energy up to 10 MeV at currents up to 60 µA. In the bremsstrahlung setup the electron beam was incident on a copper radiator, and after passing a copper collimator the induced bremsstrahlung reached the fission target placed inside the fission detector. As mentioned above, the detector was a twin Frisch grid ionization chamber, built at the Institute for Reference Materials and Measurements (IRMM) in Belgium. It has been described in detail elsewhere [6].

Two targets were used for the experiment, consisting of 130 µg/cm^2 ^{238}UF$_4$ and 190 µg/cm^2 ^{234}UF$_4$, respectively. The target material rests on a 50 µg/cm^2 gold foil with a 35 µg/cm^2 polyimide backing. The counting gas was P-10 (90% Ar + 10% CH$_4$), continuously flowing through the chamber and kept at a pressure of $1.18 \cdot 10^5$ Pa. The setup allows detection of fission fragments in an almost 4π solid angle, but for a satisfactory energy resolution a cutoff angle is introduced in the analysis. The angular distribution in low energy photofission of even-even actinide nuclei has a minimum at 0° relative to the photon momentum [7]. In order to maximize the yield in the angular cone of accepted events, the entire chamber was tilted 45° relative to the beam axis.

3. Treatment of Data

The bremsstrahlung spectra reaching the fission target were simulated with the software package Geant4 [8] for every energy of the electron beam. The average excitation energy of the compound nucleus was then derived from the convolution of the bremsstrahlung spectra and the photofission cross section [9]. The results of this calculation is summarized in Table 1, the width of the excitation energy is indicated by the standard deviation.

Table 1. Summary of investigated reactions, bremsstrahlung endpoint energy, average exciation energy and standard deviation of the excitation energy.

Reaction	E_0 (MeV)	$<E_{exc}>$ (MeV)	σ_{Eexc} (MeV)
$^{234}U(\gamma,f)$	6.8	5.77	0.43
	7.5	6.11	0.62
	9.0	6.85	0.99
$^{238}U(\gamma,f)$	6.5	5.93	0.26
	7.0	6.17	0.37
	8.5	7.06	0.73

By means of the double kinetic energy technique the information on energy and mass of the fission fragments was calculated from the pulse-height data in a well established iterative manner, as outlined in Ref. [10]. In the iterative calculation, data from Refs. [11, 12] on the average number of neutrons evaporated as function of the fragment mass was used. Also included in the iterative process is a correction for the pulse height defect in the counting gas, following a calibration procedure as suggested in Ref. [13]. The calibration was made at the highest bremsstrahlung endpoint energies measured; 8.5 MeV and 9.0 MeV for the ^{238}U and the ^{234}U experiments, respectively, using data from Refs. [14, 15]. Before the iterations can be started, systematic errors in the recorded pulse heights due to the detection process need to be accounted for. These include Frisch grid inefficiency [16] and energy loss in the sample material [6]. These effects are both angle dependent, and for correct treatment, the emission angle of the fission fragment relative to the chamber axis is needed. In this work the emission angle was extracted from the drift time of ionization electrons [17].

4. Results and Discussion

The pre-neutron mass distributions determined in this work are shown in Figure 1. Mean heavy fragment mass and width of the asymmetric mass peak from ^{238}U(γ,f) are shown as function of the average excitation energy of the compound nucleus in Figure 2. Good agreement with the results from Ref. [14] is observed. The trend of the mean heavy mass with decreasing excitation energy can be attributed to an increase in the mass yield at A $\approx$ 125-136, with a simultaneous decrease in the yield for more asymmetric masses.

Further analysis of the data was performed within the multi-modal random neck rupture model [18]. The following function was fitted to the experimental yield as a function of mass and TKE

$$Y(A, TKE) = \sum_m \frac{w_m}{\sqrt{2\pi\sigma_{A,m}^2}} \exp\left[-\frac{(A - <A>_m)^2}{2\sigma_{A,m}^2}\right] \times \left[\frac{200}{TKE}\right]^2 \times \exp\left[-\frac{(L - l_{max,m})^2}{(L - l_{min,m})l_{dec,m}}\right], \quad (2)$$

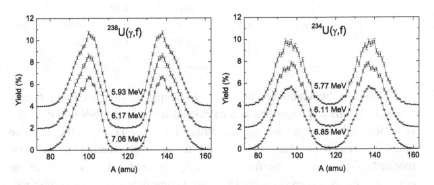

Figure 1. Pre-neutron mass distributions for ^{238}U(γ,f) (left) and ^{234}U(γ,f) (right). The average excitation energies of the compound nucleus are indicated above each distribution, which have been consecutively displaced by 2%.

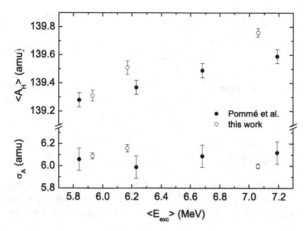

Figure 2. Mean heavy fragment mass number (upper part) and standard deviation of the asymmetric mass peak (lower part) as function of the calculated average excitation energy from this work (open circles) and from Ref. [14] (full circles).

where the sum is over the contributing fission modes. The mass yield of each mode is represented by a Gaussian centered at $<A>_m$ with a standard deviation $\sigma_{A,m}$. The variable L represents the distance between the two fragments charge centra at scission. It is parameterized from the TKE by $L = Z_L Z_H e^2/TKE \approx A_L A_H (Z_F/A_F)^2 e^2/TKE$, where the indices L, H and F represent light fragment, heavy fragment and fissioning nucleus, respectively. The parameter w_m is referred to as the fission mode weight and describes the relative yield of each mode. In fission of uranium three modes are predicted; a symmetric superlong (SL) mode, and two asymmetric standard modes (S1, S2). At excitation energies relevant for this experiment the weight of the SL mode is predicted to be in the order of 1 %. This is the same order of magnitude as the uncertainty in this parameter from the fits, and therefore the SL mode was not considered. In Figure 3 the weight of the two standard modes are shown as function of the average excitation energy of the compound nucleus. As expected, from the width of the compound nucleus excitation energy due to the nature of the bremsstrahlung, no large fluctuations in the fission mode weights are observed. The fission mode weights for ^{238}U(γ,f) are in fair agreement with literature [11]. Fission mode weights for ^{234}U(γ,f) have been extracted for the first time. The results agree well with experimental results on thermal neutron induced fission of ^{233}U [19] at an average excitation energy of 6.85 MeV.

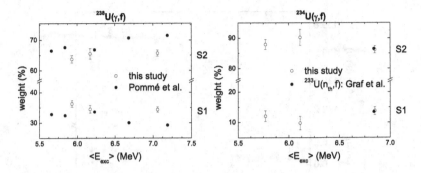

Figure 3. Weights of S1 and S2 modes as function of the average excitation energy of the compound nucleus. The plots refer to the reaction ^{238}U(γ,f) (left) from this study (open circles) and from Ref. [11] (full circles), and to ^{234}U(γ,f) (right) from this study (open circles) and for ^{233}U(n,f) from Ref. [19].

Acknowledgement

We wish to thank the S-DALINAC operating crew around R. Eichhorn for providing excellent beams. This work is supported by Deutsche Forschungsgemeinschaft through SFB 634.

References

1. G. E. Mitchell, J. D. Bowman, S. I. Penttilä, E. I. Sharapov, *Phys. Rep.* **354** (2001) 157.
2. O. Shushkov, V. Flambaum, *JETP Lett.* **32** (1980) 352.
3. G. E. Mitchell, J. D. Bowman, H. A. Weidenmüller, *Rev. Mod. Phys.* **71** (1999) 445.
4. Y. Poltoratska et al., in: *AIP Conf. Proc.*, American Institute of Physics, College Park, Maryland, USA, 2009, **1149**, pp. 983-986.
5. A. Richter, in: *Proceedings of EPAC 96*, CRC Press, Boca Raton, FL, 1996, pp. 110–114.
6. C. Budtz Jørgensen, H. H. Knitter, Ch. Straede, F.–J. Hambsch, R. Vogt, *Nucl. Instr. Meth.* **A258** (1987) 209.
7. E. Jacobs, U. Kneissl, in: C. Wagemans, *The Nuclear Fission Process*, CRC Press, Boca Raton, FL, 1991.
8. S. Agostinelli et al., *Nucl. Instr. and Meth.* **A506** (2003) 250.
9. M. Chadwick et al., *Nucl. Data Sheets* **107** (2006) 2931; http://www.nndc.bul.gov.
10. E. Birgersson, A. Oberstedt, S. Oberstedt, F.-J. Hambsch, *Nucl. Phys.* **A817** (2009) 1.

11. S. Pommé, Doctoral dissertation, Gent University (1992).
12. V. F. Apalin, Yu. N. Gritsyuk, I. E. Kutikov, V. I. Lebedev, L. A. Mikealian, Nucl. Phys. 71 (1965) 533.
13. F. -J. Hambsch, J. van Aarle, R. Vogt, Nucl. Instr. Meth. A361 (1995) 257.
14. S. Pommé et al., Nucl. Phys. A572 (1994) 237.
15. M. Verboven, E. Jacobs, M. Piessens, S. Pommé, D. De Frenne, A. De Clercq, Phys. Rev. C42 (1990) 453.
16. O. Bunemann, T. E. Cranshaw and J. A. Harvey, Can. Jour. Re. A27 (1949) 191.
17. A. Göök, M. Chernykh, J. Enders, A. Oberstedt, S. Oberstedt, Nucl. Instr. Meth. A (2010), In Press, doi:10.1016/j.nima.2010.05.009.
18. U. Brosa, S. Grossmann and A. Müller, Phys. Rep. 4 (1990) 167.
19. U. Graf, F. Gönnenwein, P. Geltenbort, K. Schreckenbach, Z. Phys. A351 (1995) 281.

STRUCTURE EFFECTS IN THE ASYMMETRIC FISSION OF ^{118}BA, ^{122}BA COMPOUND NUCLEI

G. ADEMARD*, J. P. WIELECZKO, E. BONNET, A. CHBIHI, J. D. FRANKLAND
ET AL.,
FOR THE E475S COLLABORATION
GANIL, CEA/DSN et CNRS-IN2P3,
B.P. 55027, F-14076, Caen Cedex, France
*E-mail: ademard@ganil.fr

Kinetics energies, angular distributions and cross sections of fragments with atomic number 6≤Z≤28 were measured in ^{78,82}Kr + ^{40}Ca reactions at 5.5 AMeV incident energy by means of the 4π-INDRA array at GANIL. Global features are compatible with a binary fission from compound nucleus. The cross-sections indicate the coexistence of macroscopic behaviour and structure effects that depend on the chemical composition of the emitting nucleus. Fragment-particle coincidences show that fragment with 12≤Z are excited below the separation energies of light particles.

Keywords: Decay channels; Compound nucleus; Even-odd staggering.

1. Introduction

The decay modes of warm nuclei are mainly ruled by the temperature, the angular momentum and the topology of the multidimensional space build with the relevant collective degrees of freedom. Thermal and collective properties of nuclei involve quantities that depend on the chemical composition (N/Z) of the nuclear system. In this work, we investigate the influence of the (N/Z) on binary decays in ^{78,82}Kr+^{40}Ca reactions at 5.5 A MeV incident energy.

2. Experimental results and analysis

Self-supporting 1mg/cm^2 thick ^{40}Ca targets were bombarded with 5.5 AMeV ^{78,82}Kr beams delivered at the GANIL facility. The kinetic energy and atomic number of the ejectiles were measured using the 4π-INDRA array.[1] The present analysis deals with the charged products emitted at

$3° \leq \theta_{lab} \leq 45°$ covering the forward hemisphere in the center-of-mass (c.o.m) of the reaction. In this angular range, the detection module comprises an ionization chamber (IC), a silicon detector (Si) followed with a CsI scintillator. The energy calibration was obtained using alpha particles from a Cf source and elastically scattered projectiles selected with the CIME cyclotron.

The c.o.m kinetic energy distributions of the fragments are Gaussian-like and the average values are compatible with a mechanism dominated by the Coulomb interaction. Meanwhile, the angular distributions of the fragments follow a $1/\sin\theta_{c.o.m}$ dependence. These features suggest a high degree of relaxation as expected for an emission from a compound nucleus (CN).

Fig. 1 presents the cross-sections σ_Z for fragments with atomic number $6 \leq Z \leq 28$ for the ^{78,82}Kr+^{40}Ca reactions. The σ_Z distributions exhibit a maximum around half of the total available charge. σ_Z for fragments with $Z \leq 10$ presents a strong even-odd staggering (e-o-s), and this effect persists for higher Z with a smaller amplitude. It is worth noticing that the yields around the symmetric splitting are about 30% smaller for the neutron rich system in agreement with the expectations from liquid drop model. Moreover, the influence of the (N/Z) ratio is seen in the yields of the light fragments. σ_Z for odd-Z fragments are higher for the neutron rich CN while cross-sections for even-Z fragments are higher for the neutron poor CN.

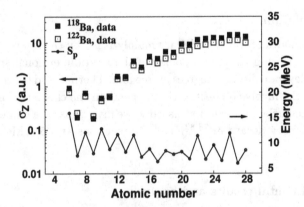

Fig. 1. Experimental cross sections for fragments emitted in ^{78}Kr+^{40}Ca (full squares) and for the ^{82}Kr+^{40}Ca (empty squares) reactions. The broken line shows the proton separation energy.

To better understand the fragment emission mechanism, we perform an event-by-event analysis of the light charged particles (LCPs) in coincidence with fragments. We first calculated, for each fragment, the relative velocities between that fragment and all light charged particles in the event. In the following step we determine the velocity components parallel ($V_\parallel$) and perpendicular ($V_\perp$) to the axis representing the direction of the fragment emission velocity in the c.o.m frame. Thus, this treatment allows to construct a common reference frame designated by Z-frame (Compl-frame) for light (heavy) fragments having different emission angle in the c.o.m frame.

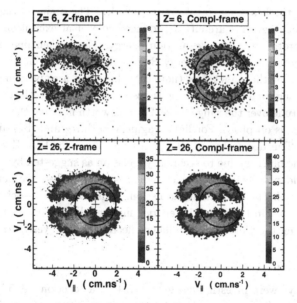

Fig. 2. $V_\parallel$-$V_\perp$ diagrams of alpha particles detected in coincidence with C (top line) and Fe (bottom line) fragments produced in ^{78}Kr+^{40}Ca reaction (see text). The velocities are calculated in the reference frames of the light fragment (left panels) and the complementary fragment (right panels)

Fig. 2 present typical examples of $V_\parallel$-$V_\perp$ diagrams for alpha particles detected in coincidence with C (top panels) and Fe (bottom panels) fragments emitted in ^{78}Kr+^{40}Ca reaction. The black circles represent the average velocities deduced from the Parker's systematic.[2] For particles emitted in coincidence with Z=6, the location of the relative velocities draw a circle centered at zero in the Compl-frame (top right panel) and no charged particles seems to be emitted by the light partner (top left panel). For Z=26, both fragments are emitters as illustrated by the presence of two regions centered at both reference frames. Thus, we observe an evolution of the light particle emission from asymmetric to symmetric fission, the change of behaviour happens around Z=12 and no circle centered at zero is present for light fragment with Z smaller than this transition value while such circle appears when Z is larger. Same conclusions hold for the coincidences with the protons and in ^{82}Kr+^{40}Ca reaction. Extensive simulations were performed in order to check that these results are not related to geometrical acceptance.

Thus, in ^{78,82}Kr+^{40}Ca reactions at 5.5 AMeV, the LCPs are emitted by both fragments in case of symmetric fission, while for a very asymmetric fission, only the heavy fragment emits particles. The main consequence is that the light fragments are either produced cold or at excitation energies below the proton or alpha separation energy. The broken line in Fig. 2 shows the proton separation energy S_p calculated for the most abundant element given by the mass tables (note that the values are given on the right scale). A strong even-odd staggering is observed for the S_p with roughly the same magnitude over the range $6 \leq Z \leq 28$. It is worth noticing that the e-o-s of S_p and σ_Z is on phase. For light fragment both σ_Z and (S_p) are larger for even Z. The decreasing of the staggering of σ_Z for high Z would be related to the blurring due to secondary emission as suggested by the study of the coincidence. Thus, the σ_Z for light fragments reflect the persistence of structural effects in asymmetric fission. This could be associated to a microscopic part in the potential energy surface which is a key ingredient in determining the yields. Such influence needs further investigations.

3. Conclusion

Binary decays were investigated in ^{78,82}Kr+^{40}Ca reactions at 5.5 A MeV incident energy. Experimental features of the fragments are compatible with a disintegration from compound nuclei. The persistence of structural effects that reflects the N/Z ratio of the nuclear system is evidenced from elemental cross-sections. Fragment-particle coincidences show that light fragments are excited below the separation energies of light particles.

References

1. J. Pouthas et al., *Nucl. Inst. Meth. Phys. Res.* **A 357**, 418 (1995).
2. W. E. Parker et al., *Phys. Rev.* **C 44**, 774 (1991).

135

CONSTRAINING FISSION PARAMETERS FOR HIGHLY EXCITED COMPOUND NUCLEI

DAVIDE MANCUSI* and JOSEPH CUGNON

University of Liège, AGO Department, allée du 6 août 17, bât. B5, B-4000 Liège 1, Belgium
E-mail: d.mancusi@ulg.ac.be

ROBERT J. CHARITY

Department of Chemistry, Washington University, St. Louis, Missouri 63130, USA

We present a statistical-model description of fission, in the framework of compound-nucleus decay, which is found to simultaneously reproduce data from both heavy-ion-induced fusion reactions and proton-induced spallation reactions around 1 GeV. For the spallation reactions, the initial compound-nucleus population is predicted by the Liège intranuclear cascade model. We are able to reproduce experimental fission probabilities in the all reactions with the same parameter set. We also discuss the need for fission transients, which are expected to have a significant effect on the spallation reactions.

Keywords: fusion reactions, spallation reactions, fission transients

1. Introduction

Although seventy years have passed since the seminal works of Bohr and Wheeler[1] and Weisskopf and Ewing[2] and the establishment of a qualitative understanding of the de-excitation mechanism of excited nuclei, quantitatively accurate and universally applicable models do not exist yet. One way to lift the degeneracy of the ingredients of the model is to explore diverse regions of the compound-nucleus parameter space. The production of excited compound nuclei can proceed from several entrance reactions. There has been a long history of compound-nucleus studies using heavy-ion-induced fusion reactions. These reactions allow one to specify the compound-nucleus mass, charge and excitation energy, however a distribution of compound-nucleus spins is obtained. Alternatively, one can consider the production of excited compound nuclei through spallation reactions. For these reactions, the need for a model to predict the initial compound-nucleus mass, charge, excitation, and spin distributions adds some uncertainty in our ability to constrain the statistical-model parameters by fitting data. However, spallation reactions allow us to explore different regions of compound-nucleus spin and excitation energy than can be probed with fusion reactions alone and thus can be important in parameter fitting.

This paper discusses the application of the GEMINI++ de-excitation model[3] to the description of fission in fusion and spallation reactions. In the latter case, the description of the entrance channel is provided by a coupling to the Liège Intranuclear Cascade model (INCL).[4] Both INCL and GEMINI++ are among the most sophisticated models in their own fields. We compare the predictions of the models with experimental residue yields in spallation studies and to fission and evaporation-residue excitation functions measured in heavy-ion induced fusion reactions.

2. The GEMINI++ and INCL4.5 models

GEMINI++ is an improved version of the GEMINI model, developed by R.J. Charity.[5] The de-excitation of the remnant proceeds through a sequence of binary decays until particle emission becomes energetically forbidden or improbable due to competition with gamma-ray emission. For fissile systems, the total fission yield is obtained from the Bohr-Wheeler formalism[1] and it competes against emission of nucleons and light nuclei. The width of the mass distribution is taken from systematics compiled by Rusanov et al.[6]

Level densities were calculated with the Fermi-gas form. The level-density parameter $a(U)$ is excitation-energy dependent with an initial fast dependence due to the washing out of shell effects following Ref. 7 and a slower dependence needed to fit the evaporation spectra. The shell-smoothed level-density parameter was assumed to have the form

$$\widetilde{a}(U) = \frac{A}{k_\infty - (k_\infty - k_0)\exp\left(-\frac{\kappa}{k_\infty - k_0}\frac{U}{A}\right)} \qquad (1)$$

which varies from A/k_0 at low excitation energies to A/k_∞ at large values. Here k_0=7.3 MeV, consistent with neutron-resonance counting data at excitation-energies near the neutron separation energy, and k_∞=12 MeV. The parameter κ defines the rate of change of $\widetilde{a}$ with energy and it is essentially zero for light nuclei (i.e. a constant $\widetilde{a}$ value) and increases roughly exponentially with A for heavier nuclei.

The INCL model[4,8] can be applied to collisions between nuclei and pions, nucleons or light nuclei of energy lower than a few GeV. The particle-nucleus collision is modelled as a sequence of binary collisions among the particles present in the system; particles that are unstable over the time scale of the collision, notably Δ resonances, are allowed to decay. The nucleus is represented by a square potential well whose radius depends on the nucleon momentum; thus, nucleons move on straight lines until they undergo a collision with another nucleon or until they reach the surface, where they escape if their total energy is positive and if they manage to penetrate the Coulomb barrier.

3. Adjustment of fission yields

Fusion and spallation reactions populate different regions of the compound-nucleus excitation-energy/spin plane. For the 1-GeV $p+^{208}$Pb spallation reaction, the INCL4.5 model predicts average values of about 167 MeV and 16.5 $\hbar$, respectively, but both distributions are quite broad and extend up to ~ 650 MeV and ~ 50 $\hbar$. On the other hand the fusion reactions we considered are characterized by higher spins and lower excitation energies. If we restrict ourselves to the regime of complete fusion ($E_{\text{beam}}/A < 10$ MeV), then we can explore somewhat higher excitation energies with more symmetric entrance channels, but high spins will still be populated. Thus it is clear that the comparison between spallation and fusion data represents a promising tool to extend the predictive power of the model over a wide region of mass, energy and spin.

For spallation reactions, we focused our efforts on proton-induced fission reactions on ^{197}Au, ^{208}Pb and ^{238}U around 1 GeV, measured in reverse kinematics with the FRagment Separator (FRS) at SIS, GSI, Darmstadt, Germany.[9-13]

3.1. *Modifications of the fission width*

The Bohr-Wheeler fission width,

$$\Gamma_{\text{BW}} = \frac{1}{2\pi\rho_n(E^*, J)} \int d\epsilon \, \rho_f(E^* - B(J) - \epsilon, J),$$

is sensitive to the choice of the fission barrier B and to the level-density parameters a_f and a_n associated with the saddle-point and ground-state configurations. The energy dependence of a_f was assumed identical for the ground state and the saddle point (Eq. (1)) and the magnitude of the two parameters was varied by a constant scaling factor a_f/a_n to account for the increased surface area of the saddle-point configuration.[14] Moreover, Lestone[15] developed a treatment of fission which explicitly includes the tilting collective degree of freedom at saddle point. We have tried to reproduce simultaneously fission cross sections from fusion and spallation experiments by (a) scaling the Sierk fission barrier by a constant factor, (b) scaling the decay width by a constant factor, (c) adjusting the a_f/a_n ratio, (d) using either the Bohr-Wheeler or the Lestone formalism and (e) introducing a constant, step-like fission delay.

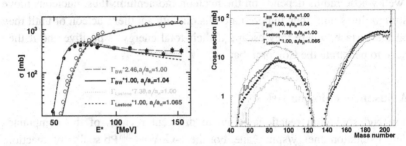

Figure 1. (Color online) Comparison of GEMINI++ predictions to the experimental evaporation-residue and fission excitation fuctions for the ^{19}F+^{181}Ta reaction (left) and to experimental residual mass distributions the for the 1-GeV p+^{208}Pb reaction (right). Experimental data from Refs. 10,16,17.

3.2. Results

Examples of fits to the ^{19}F+^{181}Ta→^{200}Pb fission and evaporation-residue excitation functions are shown in Fig. 1, left panel. As the sum of these quantities (the fusion cross section) is fixed, the degree to which the fission probability is reproduced is best gauged by the fit to the smaller quantity, i.e. σ_{fis} at low bombarding energies and σ_{ER} at the higher values. Good fits were obtained with four different parameter sets, (Fig. 1, left panel).

As it is impossible to distinguish these four ways of modifying the fission probability from the fusion data alone, we now consider the constraint of adding the spallation data to the analysis. In Fig. 1, right panel, we show the equivalent calculations for the mass distributions of the products produced in the 1-GeV p+^{208}Pb spallation reactions. Of all these possibilities, the $\Gamma_{\text{BW}} \times 1.00, a_f/a_n = 1.04$ calculation reproduces the yield of the fission peak best.

Comparison of GEMINI++ prediction to experimental fission and evaporation-residue excitations functions are shown in Fig. 2. For spallation, Fig. 3 shows the comparison between measured and calculated residue mass distributions. The central result is that it is possible to reproduce the total fission cross section for all the studied spallation reactions with only one free parameter, namely the a_f/a_n ratio.

The solution found is however unique only as long as fission delays are not considered. To show the sensitivity of predictions to these transients, we have incorporated a simple implementation of these in GEMINI++; the fission width is set to zero for a time t_{delay}, after which it assumes the Bohr-Wheeler value. During this fission-delay period, the compound nucleus can decay by light-particle evaporation and intermediate-mass-fragment emission. As Fig. 4 shows, some fission delay can be accommodated for the Lestone fission width with $a_f/a_n = 1.065$.

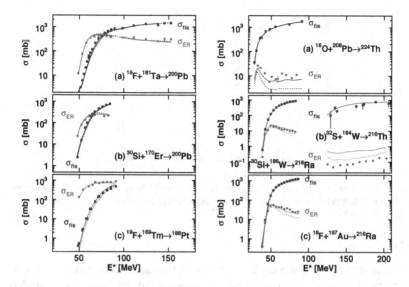

Figure 2. (Color online) Comparison of experimental and calculated fission and evaporation residue excitation function for the indicated reactions. Solid lines: Bohr-Wheeler fission width, $a_f/a_n = 1.04$, no fission delay. Dashed lines: Lestone fission width, $a_f/a_n = 1.065$, 1-zs fission delay. Experimental data from Refs. 16–24

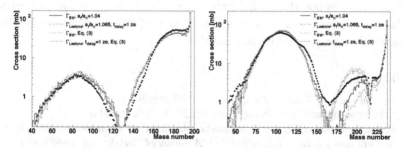

Figure 3. (Color online) Residue-mass distribution for the $p+^{197}$Au (left) and $p+^{238}$U (right) reactions at 1 GeV. Predictions of the INCL4.5-GEMINI++ code are shown for different adjustments of the fission width. Experimental data from Ref. 9,11–13.

A 1-zs fission delay approximately fits the experimental fission cross section for 1-GeV $p+^{208}$Pb. Figs. 2 and 3 also demonstrate that the agreement of the Lestone parameter set, with a constant fission delay of 1 zs, is as good as the Bohr-Wheeler parameter set, without any fission delay.

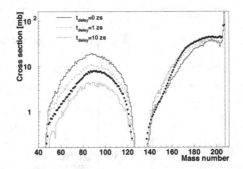

Figure 4. (Color online) Residue-mass distribution for the $p+^{208}$Pb reaction at 1 GeV. Predictions of the INCL4.5-GEMINI++ code are shown for different values of the fission delay. Experimental data from Refs. 10,16,17.

4. Conclusions

We have described the first coupling of the Liège Intranuclear Cascade model with the GEMINI++ compound-nucleus de-excitation model. We have demonstrated that it is possible to describe fission cross sections from spallation and heavy-ion fusion reactions within the same framework. The simultaneous application of the statistical-decay model to spallation and fusion actually allows us to lift some of the degeneracy of the model parameters. We were able to reproduce all the data with two different parameter sets: one that includes a small fission delay (1 zs) and one that does not. However, the existence of larger fission delays cannot be excluded. Both parameters sets are estimated to be effective for spins up to $60\hbar$ and excitation energies up to ~ 300-400 MeV.

Bibliography

1. N. Bohr and J. A. Wheeler, *Phys. Rev.* **56**, 426(September 1939).
2. V. F. Weisskopf and D. H. Ewing, *Phys. Rev.* **57**, 472(March 1940).
3. R. J. Charity, GEMINI: a code to simulate the decay od a compound nucleus by a series of binary decays, in *Joint ICTP-IAEA Advanced Workshop on Model Codes for Spallation Reactions*, (Trieste, Italy, 2008). Report INDC(NDC)-0530.
4. A. Boudard, J. Cugnon, S. Leray and C. Volant, *Phys. Rev. C* **66**, p. 044615(October 2002).
5. R. J. Charity, M. A. McMahan, G. J. Wozniak, R. J. McDonald, L. G. Moretto, D. G. Sarantites, L. G. Sobotka, G. Guarino, A. Pantaleo, L. Fiore, A. Gobbi and K. D. Hildenbrand, *Nucl. Phys. A* **483**, 371(June 1988).
6. A. Y. Rusanov, M. G. Itkis and V. N. Okolovich, *Phys. Atom. Nucl.* **60**, 683(May 1997).
7. A. V. Ignatyuk, G. N. Smirenkin and A. S. Tishin, *Sov. J. Nucl. Phys.* **21**, p. 255 (1975).
8. J. Cugnon, A. Boudard, S. Leray and D. Mancusi. in preparation, (2010).
9. J. Benlliure, P. Armbruster, M. Bernas, A. Boudard, J.-P. Dufour, T. Enqvist,

R. Legrain, S. Leray, B. Mustapha, F. Rejmund, K.-H. Schmidt, C. Stéphan, L. Tassan-Got and C. Volant, *Nucl. Phys. A* **683**, 513(February 2001).

10. T. Enqvist, W. Wlazło, P. Armbruster, J. Benlliure, M. Bernas, A. Boudard, S. Czajkowski, R. Legrain, S. Leray, B. Mustapha, M. Pravikoff, F. Rejmund, K.-H. Schmidt, C. Stéphan, J. Taïeb, L. Tassan-Got and C. Volant, *Nucl. Phys. A* **686**, 481(April 2001).

11. M. Bernas, P. Armbruster, J. Benlliure, A. Boudard, E. Casarejos, S. Czajkowski, T. Enqvist, R. Legrain, S. Leray, B. Mustapha, P. Napolitani, J. Pereira, F. Rejmund, M. V. Ricciardi, K.-H. Schmidt, C. Stéphan, J. Taïeb, L. Tassan-Got and C. Volant, *Nucl. Phys. A* **725**, 213(September 2003).

12. J. Taïeb, K.-H. Schmidt, L. Tassan-Got, P. Armbruster, J. Benlliure, M. Bernas, A. Boudard, E. Casarejos, S. Czajkowski, T. Enqvist, R. Legrain, S. Leray, B. Mustapha, M. Pravikoff, F. Rejmund, C. Stéphan, C. Volant and W. Wlazło, *Nucl. Phys. A* **724**, 413(September 2003).

13. M. V. Ricciardi, P. Armbruster, J. Benlliure, M. Bernas, A. Boudard, S. Czajkowski, T. Enqvist, A. Kelić, S. Leray, R. Legrain, B. Mustapha, J. Pereira, F. Rejmund, K.-H. Schmidt, C. Stéphan, L. Tassan-Got, C. Volant and O. Yordanov, *Phys. Rev. C* **73**, p. 014607(January 2006).

14. J. Tōke and W. Światecki, *Nucl. Phys. A* **372**, p. 141 (1981).

15. J. P. Lestone, *Phys. Rev. C* **59**, 1540(March 1999).

16. D. J. Hinde, J. R. Leigh, J. O. Newton, W. Galster and S. Sie, *Nucl. Phys. A* **385**, p. 109 (1982).

17. A. L. Caraley, B. P. Henry, J. P. Lestone and R. Vandenbosch, *Phys. Rev. C* **62**, p. 054612(Oct 2000).

18. R. J. Charity, L. G. Sobotka, J. F. Dempsey, M. Devlin, S. Komarov, D. G. Sarantites, A. L. Caraley, R. T. deSouza, W. Loveland, D. Peterson, B. B. Back, C. N. Davids and D. Seweryniak, *Phys. Rev. C* **67**, p. 044611(Apr 2003).

19. K.-T. Brinkmann, A. L. Caraley, B. J. Fineman, N. Gan, J. Velkovska and R. L. McGrath, *Phys. Rev. C* **50**, 309(Jul 1994).

20. F. Videbæk, R. B. Goldstein, L. Grodzins, S. G. Steadman, T. A. Belote and J. D. Garrett, *Phys. Rev. C* **15**, 954(Mar 1977).

21. B. B. Back, R. R. Betts, J. E. Gindler, B. D. Wilkins, S. Saini, M. B. Tsang, C. K. Gelbke, W. G. Lynch, M. A. McMahan and P. A. Baisden, *Phys. Rev. C* **32**, 195(July 1985).

22. B. B. Back, D. J. Blumenthal, C. N. Davids, D. J. Henderson, R. Hermann, D. J. Hofman, C. L. Jiang, H. T. Penttilä and A. H. Wuosmaa, *Phys. Rev. C* **60**, p. 044602(August 1999).

23. J. G. Keller, B. B. Back, B. G. Glagola, D. Henderson, S. B. Kaufman, S. J. Sanders, R. H. Siemssen, F. Videbaek, B. D. Wilkins and A. Worsham, *Phys. Rev. C* **36**, 1364(Oct 1987).

24. A. C. Berriman, D. J. Hinde, M. Dasgupta, C. R. Morton, R. D. Butt and J. O. Newton, *Nature (London)* **413**, p. 144 (2001).

Ternary Fission

145

CHARACTERISTICS OF LIGHT CHARGED PARTICLE EMISSION IN THE TERNARY FISSION OF ^{250}CF AND ^{252}CF AT DIFFERENT EXCITATION ENERGIES

S. VERMOTE[†] AND C. WAGEMANS
Department of Physics and Astronomy, University of Gent, B-9000 Gent, Belgium

O. SEROT
CEA Cadarache, DEN/DER/SPRC/LEPh, F-13108 Saint-Paul-lez-Durance, France

J. HEYSE
*SCK•CEN, Boeretang 200, B-2400 Mol, Belgium and
EC-JRC Institute for Reference Materials and Measurements, B-2440 Geel, Belgium*

T. SOLDNER[*] AND P. GELTENBORT
Institut Laue-Langevin, F-38042 Grenoble, France

I. ALMAHAMID
Wadsworth Center, New York State Department of Health, Albany, NY 12201, USA

G. TIAN AND L. RAO
Lawrence Berkeley National Laboratory, Berkeley, CA 94720, USA

The emission probabilities and the energy distributions of tritons, α and ^{6}He particles emitted in the spontaneous ternary fission (zero excitation energy) of ^{250}Cf and ^{252}Cf and in the cold neutron induced fission (excitation energy ≈ 6.5 MeV) of ^{249}Cf and ^{251}Cf are determined. The particle identification was done with suited ΔE-E telescope detectors, at the IRMM (Geel, Belgium) for the spontaneous fission and at the ILL (Grenoble, France) for the neutron induced fission measurements. Hence particle emission characteristics of the fissioning systems ^{250}Cf and ^{252}Cf are obtained at zero and at about 6.5 MeV excitation energies. While the triton emission probability is hardly influenced by the excitation energy, the ^{4}He and ^{6}He emission probability in spontaneous fission is higher than for neutron induced fission. This can be explained by the strong influence of the cluster preformation probability on the ternary particle emission probability.

[†] Corresponding author. E-mail address: sofie.vermote@ugent.be.
[*] Presently at Physics Department E18, TU Munich, 85748 Garching, Germany.

1. Introduction

Nuclear fission is generally a binary process. However, once every 300-400 fission events a light charged particle accompanies the two fission fragments. This process is called ternary fission.

We will focus on the characteristics of the particles with the largest emission probability, i.e. α particles (also called Long Range Alpha or LRA particles), tritons (t) and ^{6}He particles, for two reasons: (a) triton emission yields are needed by nuclear industry for safe manipulations of radioactive waste [1]; (b) since a different behavior of ternary α and triton emission was observed previously [2], we wanted to answer the question of how ^{6}He particles will behave.

Californium isotopes provide an excellent opportunity to investigate the influence of the cluster preformation probability factor, as well as the influence of the excitation energy of the compound nucleus. These effects are studied here by measuring the fissioning systems ^{250}Cf and ^{252}Cf at zero excitation energy (spontaneous fission) and at an excitation energy of about 6.5 MeV (neutron induced fission).

This paper gives an overview of the results of the study of Cf isotopes. More details can be found in [3].

2. Experimental setup

The spontaneous fission of ^{250,252}Cf has been studied at the Institute for Reference Materials and Measurements (IRMM) in Geel, Belgium. The ^{249,251}Cf neutron induced fission measurements were carried out at the PF1b cold neutron beam facility installed at the High Flux Reactor of the Institut Laue-Langevin (ILL) in Grenoble, France.

2.1. Sample characteristics

Highly enriched ^{249}Cf and ^{250}Cf samples were prepared at the Lawrence Berkeley National Laboratory in the USA. Both ^{251}Cf and ^{252}Cf samples were prepared at the Institute of Nuclear Chemistry of Mainz University in Germany. Special attention has to be given to the isotopic composition of the ^{251}Cf sample: ^{249}Cf (17.65%), ^{250}Cf (35.40%), ^{251}Cf (46.18%) and ^{252}Cf (0.77%). Due to this composition, the spontaneous fission yield and the contribution of ^{249}Cf(n,f) were not negligible.

2.2. Detection system

For the ^{249,251}Cf neutron induced fission measurements the sample was placed in the centre of a vacuum chamber at an angle of 45 degrees with the incoming neutron beam. For the spontaneous fission of ^{250,252}Cf the same setup was used, however here we measured without neutron beam, so the sample could be placed right in front of the detectors.

The measurements were performed in two separate steps. In a first step, ternary particles were detected, allowing the determination of both energy distributions and counting rates. Therefore well-calibrated silicon surface barrier detectors were used.

In addition, ΔE detectors were covered with thin aluminum foils of 25 or 30 μm to stop α decay particles and fission fragments from penetrating the detector.

For all experiments the detector characteristics were chosen in order to have the best setup for detecting α and ^{6}He particles (ΔE detector with a thickness between 29.8 μm and 35 μm, E detector with a thickness of 500 μm), or for detecting α particles and tritons (ΔE detector with a thickness between 41 μm and 62.9 μm, E detector with a thickness of 1500 μm).

In a second step, binary fission fragments were detected in order to determine the Binary Fission Yield (B). At this stage, the ΔE detector from the telescope suited to measure LRA/B, was removed, together with the aluminum foil, and replaced by a dummy ring with exactly the same dimensions. In this way, binary fission fragments could be measured with the E-detector (which was always thick enough) under the same detection geometry as ternary particles.

3. Measurements and results

3.1. Particle identification

The procedure used to identify various ternary particles and separate them from the background is the one proposed by Goulding et al. [4]. This method is based on the difference in energy loss of different particles in the same material using the equation: $T/a = (E + \Delta E)^{1.73} - E^{1.73}$, where T is the thickness of the ΔE detector and a is a particle and material specific constant.

The selection of ternary particles was realized by putting a window on the region of interest of the T/a spectrum. In the case of the tritons, an additional correction due to the background was needed. After the selection, ΔE and E spectra were obtained for a given ternary particle and the total energy distribution could be deduced. The thresholds in energy for each ternary particle are due to the thickness of the ΔE detector, the electronic noise and the presence of the Al-foil. The average energy and the Full Width at Half Maximum (FWHM) of the energy distribution were obtained from a Gaussian fit performed on the experimental data.

3.2. Results

3.2.1. Binary fission

A typical binary fission spectrum is shown in Fig. 1. The alpha pile-up peak in the lowest channels due to the radioactive decay of the Cf isotope has to be removed. Then the remaining spectrum is extrapolated and the corresponding number of binary fission events can be deduced after integration of the extrapolated spectrum.

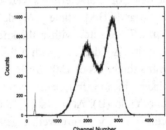

Figure 1. Measured binary fission spectrum for ^{249}Cf(n,f).

3.2.2. Ternary fission

Fig. 2 shows the spectra for the LRA, triton and ^{6}He measurements for ^{249}Cf(n,f). The characteristics of the energy distributions for the four Cf isotopes are given in Table 1. Emission probabilities relative to LRA particles are reported in Table 2 together with the absolute emission probabilities. All the uncertainties given correspond to the sum of statistical and systematic uncertainties.

Table 1. Values for average energy (E) and full width at half maximum (FWHM) for the various ternary particles measured.

	LRA		Tritons		^{6}He	
	E [MeV]	FWHM[MeV]	E [MeV]	FWHM[MeV]	E [MeV]	FWHM[MeV]
^{249}Cf	16.09 ± 0.18	10.64 ± 0.27	8.47 ± 0.19	8.52 ± 0.34	10.99 ± 0.32	10.35 ± 0.60
^{251}Cf	15.89 ± 0.12	10.60 ± 0.18	8.53 ± 0.12	8.39 ± 0.19	10.84 ± 0.36	9.98 ± 0.53
^{250}Cf	15.95 ± 0.13	10.49 ± 0.16	8.31 ± 0.30	8.58 ± 0.49	10.64 ± 0.30	10.49 ± 0.54
^{252}Cf	15.96 ± 0.09	10.22 ± 0.18	8.55 ± 0.28	8.26 ± 0.43	11.22 ± 0.52	8.95 ± 0.81

Table 2. Values for relative and absolute emission probabilities (per fission) for the various ternary particles measured.

	LRA/B [10^{-3}]	t/B [10^{-4}]	^{6}He/B [10^{-5}]	t/LRA [%]	^{6}He/LRA [%]
^{249}Cf	2.77 ± 0.11	2.13 ± 0.15	6.99 ±0.66	7.76 ± 0.50	2.54 ± 0.23
^{251}Cf	2.41 ± 0.14	2.20 ± 0.14	7.58 ±0.69	9.02 ± 0.58	3.15 ± 0.34
^{250}Cf	2.93 ± 0.10	2.08 ± 0.27	8.03 ±1.00	6.96 ± 0.89	2.74 ± 0.33
^{252}Cf	2.56 ± 0.07	1.89 ± 0.19	7.68 ±0.72	7.37 ± 0.72	3.00 ± 0.27

For LRA particles, a Gaussian fit was performed on experimental data with an energy above 12.5 MeV. In the case of tritons and ^{6}He particles, a Gaussian fit to all data points is performed.

Due to the isotopic composition of the ^{251}Cf sample, two measurements had to be performed in each step: one with the neutron beam open, measuring both the neutron induced fission and the spontaneous fission for all isotopes present in the sample, and one with closed neutron beam, to determine the contribution of the spontaneous fission of ^{250,252}Cf. Thus, results can be derived for the neutron induced fission only. However, a correction still has to be made for the ^{249}Cf(n,f) contribution. This can be done easily since experimental results with the ^{249}Cf sample are available.

In addition, for ^{252}Cf, a measurement without protective Al-foil before the ΔE detector was performed in order to examine the non-Gaussian tailing on the low-energy side of the energy distribution for the α particles. This measurement provided an α energy distribution with a low detection limit of 7.5 MeV (Fig. 3). Our results nicely agree with the non-Gaussian low-energy tailing observed by Tischenko et al. [5], confirmed also by Mutterer et al. [6].

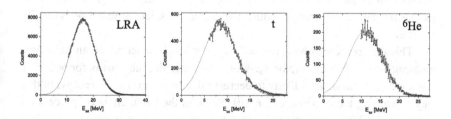

Figure 2. Energy distributions for LRA, tritons and ^{6}He for ^{249}Cf(n,f).

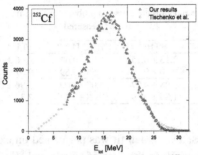

Figure 3. Energy distribution for the ternary α particles emitted in ^{252}Cf(SF), without Al protecting foil, in comparison with the results obtained by Tischenko et al. [5].

4. Discussion

In order to search for systematic trends, we base the discussion on the californium data reported above and on similar data obtained by our research group for curium isotopes [7], for the spontaneous fission of plutonium isotopes [8] and for the thermal neutron induced fission of ^{233}U, ^{235}U, ^{239}Pu, ^{241}Pu [9], ^{237}Np [10] and ^{229}Th, ^{241}Am, ^{243}Am [11].

To examine the influence of the excitation energy on the emission probabilities, Fig. 4 is shown. In the upper left figure, the LRA emission probability is plotted as a function of the fissility parameter Z^2/A of the compound nuclei. The same is done for the tritons (Fig. 4, lower part). These figures allow several observations. First, the general trend is demonstrated that both α and triton emission probabilities increase with increasing fissility, although still strong fluctuations are seen for the α particles. Another observation is that we see a decrease of the LRA emission probability with increasing excitation energy, while the triton emission probability is hardly affected.

This difference can be explained by the strong impact of the alpha cluster preformation probability factor S_α. When the fissioning nucleus is formed after capture of a neutron, S_α is likely to decrease due to the excitation energy, in this way explaining the decrease of LRA/B. Since in the triton emission process no cluster preformation is involved, a similar effect does not occur here.

A new plot (Fig. 4, upper part, right) is made for the LRA particles, showing (LRA/B)/S_α as a function of Z^2/A. In this figure, the strong fluctuations, shown in the left part of Fig. 4, are mostly gone, and the data vary now in a more smooth way as a function of Z^2/A as they do for tritons.

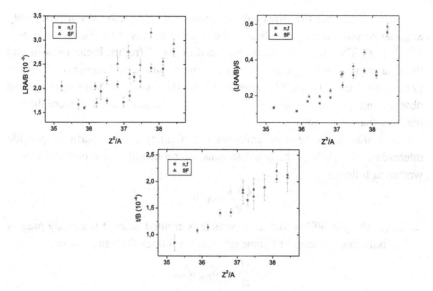

Figure 4. LRA/B (upper part, left), (LRA/B)/S_α (upper part, right) and t/B (lower part) as a function of Z^2/A of the compound nucleus.

Examining the behavior of ^{6}He particles, Fig. 5 (left) shows the absolute emission probability ^{6}He/B as a function of Z^2/A. Three isotope couples are plotted, namely ^{243}Cm(n,f) – ^{244}Cm(SF), ^{249}Cf(n,f) – ^{250}Cf(SF) and ^{251}Cf(n,f) – ^{252}Cf(SF). Again an indication of an increase of ^{6}He/B with increasing fissility is demonstrated, and in all cases a (slightly) higher value for spontaneous fission than for neutron induced fission can be observed.

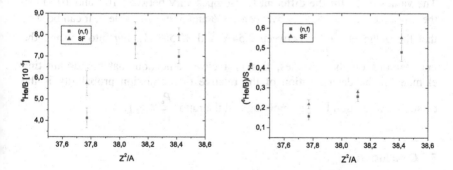

Figure 5. The absolute emission probabilities for ternary ^{6}He particles (left) and (^{6}He/B)/$S^{6\text{He}}$ (right) as a function of Z^2/A of the compound nucleus.

In analogy with the ternary α emission, a cluster preformation probability S^6_{He} can be introduced, according to the relation proposed by Blendowske: $S^6_{He} = S_\alpha^{5/3}$ [12]. The effect of S^6_{He} is illustrated in Fig. 5 (right). It can be seen that the data suggest now to vary in a more smooth way as a function of Z^2/A, as previously observed for LRA, however further data would be helpful. The above observations permit to conclude that the ^{6}He particles behave more like α particles than like tritons.

In addition, the relative emission probabilities of ^{6}He particles provide interesting information. The absolute emission probability for α particles can be written as follows:

$$\frac{LRA}{B} = S_\alpha \cdot P_{LRA}$$

with P_{LRA} the probability that an α particle is emitted when it is already present in the fissioning nucleus. The same relation is valid for the ^{6}He particles:

$$\frac{^6He}{B} = S_{^6He} \cdot P_{^6He}$$

With P^6_{He} the probability that a ^{6}He particle is emitted when it is already present in the fissioning nucleus. Dividing both relations leads to:

$$\frac{^6He}{LRA} = \frac{S_{^6He}}{S_\alpha} \cdot \frac{P_{^6He}}{P_{LRA}}$$

Taking into account the above mentioned relation by Blendowske [12], we obtain:

$$\frac{^6He}{LRA} = S_\alpha^{2/3} \cdot \frac{P_{^6He}}{P_{LRA}}$$

The values of S_α for the different Cf isotopes vary between 10^{-2} and 10^{-3} [13], therefore the value of $S_\alpha^{2/3}$ is between 2.9% and 3.9%. In Table 2, it can be seen that the value of $\frac{^6He}{LRA}$ is between 2.54% and 3.15%. Taking into account the uncertainties on these values, this is a clear indication that the dominating element in the determination of the relative 6He emission probability is the cluster preformation factor, implying that the ratio $\frac{P_{^6He}}{P_{LRA}} \cong 1$.

5. Conclusion

The present work provides experimental data on the ternary α, triton and ^{6}He emission for the fissioning systems ^{250}Cf and ^{252}Cf in the ground state and at an excitation energy of about 6.5 MeV. These results significantly enlarge the available ternary fission data base. Furthermore, this work puts into evidence the strong impact of particle preformation on the ternary particle emission probability. We demonstrate first evidence that the emission probabilities for ^{6}He particles can be described by a preformation factor S^6_{He}. This indicates that the ^{6}He particles behave more like α particles than like tritons.

Acknowledgments

Part of this work was performed in the framework of the Interuniversity Attraction Poles project financed by the Belgian Science Policy Office of the Belgian State.

References

1. R. Mills, Compilation and evaluation of fission yield nuclear data, IAEA-TECDOC-1168 (2000).
2. C. Wagemans, Proc. Seminar on Fission Pont d'Oye II, Habay-la-Neuve, Belgium, 61 (1991).
3. S. Vermote et al., *Nucl. Phys.* **A837**, 176 (2010).
4. F. Goulding, D. Landis, J. Cerny and R. Pehl, *Nucl. Instr. Meth.* **31**, 1 (1964).
5. V. Tischenko, U. Jahnke, C. Herbach and D. Hilscher, Report HMI-B 588 (2002).
6. M. Mutterer et al., *Phys. Rev.* **C78**, 064616 (2008).
7. S. Vermote et al., *Nucl. Phys.* **A806**, 1 (2008).
8. O. Serot and C. Wagemans, *Nucl. Phys.* **A641**, 34 (1998).
9. C. Wagemans, P. D'hondt, P. Schillebeeckx and R. Brissot, *Phys. Rev.* **C33**, 943 (1986).
10. C. Wagemans et al., *Nucl. Phys.* **A369**, 1, 1981.
11. C. Wagemans, Proc. Int. Workshop on Dynamical Aspects of Nuclear Fission, Smolenice, Slovak Rep., E.V. Ivashkevich, 139 (1991).
12. R. Blendowske, T. Fliesbach and H. Walliser, *Z. Phys.* **A339**, 121 (1991).
13. S. Vermote, Ph.D. Thesis, University of Gent (2009).

TERNARY FISSION STUDIES BY CORRELATION MEASUREMENTS WITH TERNARY PARTICLES

MANFRED MUTTERER *

Institut für Kernphysik, Technische Universität, 64289 Darmstadt, Germany
and
GSI Helmholtzzentrum für Schwerionenphysik GmbH, 64291 Darmstadt, Germany
* E-mail: mutterer@email.de

The rare ternary fission process has been studied mainly by inclusive measurements of the energy distributions and fractional yields of the light charged particles (LCPs) from fission, or by experiments on the angular and energy correlation between LCPs and fission fragments (FFs). The present contribution presents a brief overview of more elaborate correlation measurements that comprise the emission of neutrons and γ rays with LCPs and FFs, or the coincident registration of two LCPs. These measurements have permitted identification of new modes of particle-accompanied fission, such as the population of excited states in LCPs, the formation of neutron-unstable nuclei as short-lived intermediate LCPs, as well as the sequential decay of particle-unstable LCPs and quaternary fission. Furthermore, the neutron multiplicity numbers $\bar{\nu}(A)$ and distributions of fragment masses A, measured for the ternary fission modes with various LCP isotopes, give a valuable hint of the role played by nuclear shell structure in the fission process near scission. Finally, two different hitherto unknown asymmetries in ternary α-particle emission with respect to the fission axis, called the TRI and ROT effect, were studied in fission reactions induced by polarised cold neutrons.

Keywords: Ternary Fission, correlation studies, reactions: ^{252}Cf(sf),^{233}U(n,f), ^{235}U(n,f), ^{239}Pu(n,f), TRI and ROT effects

1. Introduction

In the ternary fission (TF) process, nucleons from the neck joining the fragments right at scission cluster into light charged particles (LCPs) that are subsequently ejected at about right angle with respect to the fission axis (see reviews[1,2]). Although TF is a rare process ($\approx 1/260$ relative to binary fission, for ^{252}Cf), and is consequently difficult to measure, it provides one of the few means to the experimentalist to explore the fissioning system near the instant of rupture.[3] Besides inclusive measurements of LCP yields and energy spectra,[4] further experiments were performed on the angular and energy correlations between LCPs and FFs, aiming to achieve a complete kinematical description of the three-body break-up, mainly for the relatively abundant α-TF mode contributing 87% of TF events (see Ref.[2] for an overview). As an example, the correlation of α energy and emission angle for ^{252}Cf was measured precisely using the detection system DIOGENES.[5] In the data analysis for a recent α-FF coincidence experiment, the DIOGENES data were used for providing reference ternary α spectra in ^{252}Cf valid for different experimental geometries.[6]

In the last two decades, a number of elaborate correlation experiments were performed which either include the registration of neutrons and γ-rays with LCPs and FFs, or the coincident registration of two LCPs, which were able to considerably enlarge our knowledge on the TF process. In the following, we are going to present a brief survey of the most prominent features of these experimental studies.

2. Experiment with the Darmstadt-Heidelberg Crystal Ball Spectrometer

2.1. *Energy Correlations in Ternary Fission*

In a first experiment,[7] the Darmstadt-Heidelberg 4π-NaI(Tl) Crystal Ball spectrometer (CB) was applied for measuring fission γ-rays[8] (with ≥ 90% efficiency) and fission neutrons[9] (with ≃ 60% efficiency). The ^{252}Cf sample and the detector system "CODIS" for the FFs and LCPs were mounted at the center of the CB. As a specific feature of the CODIS chamber the cathode is segmented into 8 sectors which permits measuring not only the energies of both FFs but also their polar and azimuthal angles of emission relative to the chamber axis. The set of measured parameters has allowed to determine, for each fission event, the following quantities and their mutual correlations: fragment masses and kinetic energies; multiplicity and angular distribution of fission neutrons; multiplicity, energy and angular distributions of fission γ-rays; energy, nuclear charge (mass), and emission angle of the LCP from ternary fission. Important results were obtained on the angular anisotropy of γ-rays relative to the fission axis for both, binary and ternary fission. The comparison of anisotropy parameters in the two processes gives valuable insight into the process of fragment spin formation in fission.[8]

From the kinematical data and the multiplicity of emitted neutrons the fragment total excitation energies TXE could be deduced, for the first time

Fig. 1. Experimental setup for studying ternary fission in ^{252}Cf(sf). The Darmstadt-Heidelberg 4π-NaI(Tl) Crystal Ball spectrometer (CB) for neutron and γ-ray registration surrounds the detector system CODIS for the measurement of fission fragments and LCPs. The ^{252}Cf sample is at the center of both detection systems. The picture shows the CB with its two halves being moved apart.

for various ternary fission modes with LCPs up to carbon. It turns out that LCP emission proceeds in expense of a considerable amount of TXE (35 MeV, on average, for binary ^{252}Cf fission), with the required energy for particle emission increasing with LCP mass and energy. As an example, the average TXE decreases from 27 MeV to 15 MeV when instead of an α-particle a ternary C-isotope is emitted. In a sense, TF with emission of heavier LCPs features a rather cold large-scale rearrangement of nuclear matter. The neutron multiplicity numbers $\bar{\nu}(A)$ and distributions of fragment masses A, measured for the ternary fission modes with various LCP isotopes, from α-TF to C-TF, give a clear evidence for a pronounced preformation of the FFs right at scission dominated by the well-known double-magic shells, mainly Z = 50 and N = 82. At scission, the biggest amount of TXE is obviously stored in the deformation due to neck formation. In TF, part of this TXE is consumed by LCP emission.[3,10]

2.2. Intermediate ^5He, ^7He and $^8Li^*$ LCPs and Yield Systematics of He Isotopes

In the same experiment, the neutron-unstable odd-N isotopes ^{5}He, ^{7}He and ^{8}Li* (in its excited state of $E^* = 2.26$ MeV) were identified to show up as intermediate LCPs in TF of ^{252}Cf.[9] The emergence of the ternary ^{5}He and ^{7}He particles (lifetimes: $1 \times 10^{-21} s$, and $4 \times 10^{-21} s$, respectively) was disclosed from the measured angular distributions of their decay neutrons focused by the emission in flight towards the direction of motion of ^{4}He and ^{6}He particles. Previously, only ternary ^{5}He emission was observed by analyzing relative neutron intensities measured at forward and backward angles with respect to the α particles.[11,12] In the present work, neutrons were observed to be peaked also around Li-particle motion, which is attributed to the decay of the second excited state at $E^* = 2.26$ MeV (lifetime: 2×10^{-20} s) of ^{8}Li. The population of ^{8}Li* was deduced to be 0.06(2), relative to Li ternary fission, and 0.33(20) relative to the yield of particle stable ^{8}Li.

The fractional yields of the ^{5}He and ^{7}He TF modes relative to "true" ternary ^{4}He and ^{6}He TF, respectively, were determined to be 0.21(5) for both cases. The mean energy of the ^{4}He residues resulting from the ^{5}He decay was determined to be 12.4(3) MeV, compared to 15.7(2) MeV for all ternary α-particles registered, and to 16.4(3) MeV for the true ternary α-particles. The mean energy of the ^{6}He residues from the ^{7}He decay is 11.0(15) MeV, compared to 12.3(5) MeV for all ternary ^{6}He particles. We note that ^{5}He in ^{252}Cf fission has the second highest yield among all LCPs, being only superseded (by a factor of $\simeq 5$) by ^{4}He emission, but downgrading ^{3}H (by a factor of $\simeq 2$) to the third most-abundant LCP. The surprisingly high yields for these exotic clusters indicate that they are formed as the stable species inside the parent nuclei.[9] However, established theoretical yield estimations,[13–15] when applied to the particle-unstable LCP species, predict about a factor of 4 lower ^{5}He and ^{7}He yields as were actually measured. This inconsistency is removed when besides the energetics at scission also the spins of the LCPs are considered in calculating LCP yields. In a systematic statistical approach, theoretical yields should be multiplied with the statistical weight factor, $(2I_i + 1)$, with I_i being the spin of the LCPs in states i.[16] With this ansatz, theoretical yield ratios ^{5}He/^{4}He, and ^{7}He/^{6}He will increase by a factor of 4 due to the $3/2^-$ spin of the ^{5,7}He ground states as compared to the 0^+ spin of ^{4,6}He. Thus, the consideration of the multiplicity of LCP states brings the experimental data into close agreement with yield calculations.

It is worthwhile to note that ternary fission with the emission of neutron-unstable LCPs provides a source of neutrons that are emitted at about right angles to the fission axis, the dominant part coming from ^{5}He, with about one neutron in every 1500 binary fission events. This source of neutrons thus may mimic the search for so-called scission neutrons[17] thought to be related to the binary fission process.

3. Correlation Experiments Performed at GSI and the ILL

We finally deal with two more recent experimental studies in the series of FF-LCP correlation experiments that either include registration of γ-rays with LCPs and FFs, or the coincident registration of two LCPs.

In a first experiment on ^{252}Cf, besides measuring LCPs and the pair of FFs, the γ-rays emitted in ternary fission have been intercepted with high-efficiency Ge detectors.[18] Of interest is here again the angular anisotropy of γ-rays relative to the fission axis for both, binary and ternary fission. But, besides γ-rays from the FFs, also the LCPs may be produced in excited states and emit γ-rays. The relative probability for LCPs to be born either in the ground or excited state may give a clue to excitation mechanisms or nuclear temperature at scission. The study of FF-LCP correlations was also the main purpose of the ^{235}U(n,f) experiment performed at the high-flux reactor of the ILL, Grenoble, using an intense cold neutron beam of 3×10^9 neutrons/(cm^{2}s) [9]. Here, the high counting rate of 2.5×10^5 fissions/s permitted to register, for the first time, FF-LCP correlations also for very rare quaternary fission events[20]

A schematic drawing of the detector setup for the ^{252}Cf experiment is given in the upper part of Fig. 2. A californium source with 2.5×10^4 fissions/s on a thin backing is placed at the center of the cathode of a twin back-to-back ionization chamber. For identifying the LCPs two rings with 12 ΔE-E$_{res}$ telescopes each are installed in the chamber, denoted as CODIS2. They are designed from ΔE ionization chambers and E$_{res}$ silicon detectors for identifying the LCPs emitted in coincidence with the FFs and determining their energies and emission angles. Further, for detecting the γ-rays emitted by FFs and LCPs two segmented large-volume Super-Clover Ge-detectors were aligned on the chamber axis in close geometry. Each detector has 4 large Ge crystals, 14 cm in length and 6 cm in diameter. More details are given in Ref. [18] In particular, due to the high resolution of the ΔE-E$_{res}$ telescopes in CODIS2, the isotopic separation up to Be-LCPs could be achieved and the yields determined.[19] For ^{252}Cf(sf), isotopic yields of Be-LCPS were hitherto not known.

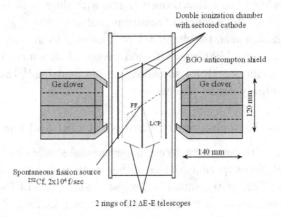

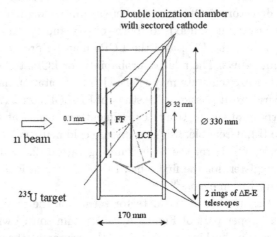

Fig. 2. Upper panel: Detector arrangement for the study of ternary fission in the spontaneous decay of ^{252}Cf(sf). The central part is the FF and LCP detector system CODIS2. The two segmented Super Glover Ge detectors on both sides of CODIS2 are equipped with BGO anticompton shields. Lower panel: Detector system CODIS2 adapted for the measurement of $^{236}U^*$ ternary fission following capture of cold neutrons.

The detector assembly for the experiment on ternary fission of $^{236}U^*$ following capture of cold neutrons is shown in the lower part of Fig. 2. A thin (50 µg/cm^2) and highly enriched ^{235}U target was placed in a cold neutron flux of 3×10^9 neutrons /(cm^{2}s). The same double ionization chamber CODIS2 as in the experiment with ^{252}Cf was used, but modifications had to be made to adapt it to the high neutron flux. The gas was CF$_4$ at 300 mb pressure, the anodes plates were replaced by grids, and the neutron entrance and exit windows were made of 0.1 mm thick Al foils. A count rate capability of 2.5×10^5/s was achieved for binary fission without compromising the overall performance. The counting rate for ternary fission was 70/s and for quaternary fission 7×10^{-4}/s. More details are given in Ref.[20]

4. Ternary Fission as a Tool to Study Fission of Rotating Nuclei

Fission reactions induced by polarised cold neutrons have so far mainly been used to study symmetry laws in fission like e.g. parity conservation.[21] Subsequently, a hitherto unknown asymmetry in ternary α-particle emission was observed by applying the neutron spin flip method to explore small changes in the α-particle angular distribution with respect to the fission axis and the neutron spin direction. The effect was found in the reactions ^{233}U(n,f) and ^{239}Pu(n,f) induced by polarised cold neutrons from the PF1 beam line installed at the high-flux reactor of the ILL, Genoble, France. It was called the "TRI" effect.[22] The asymmetry could possibly be attributed to the influence of the Coriolis force present in a rotating nucleus which is about to undergo fission and which, besides the two main fragments, is ejecting an α-particle.

In continuation of these experiments for the ^{235}U(n,f) reaction, surprisingly a startling new phenomenon, namely a shift in the angular distributions of the α-particles when flipping the neutron spin was observed. The phenomenon was dubbed the "ROT" effect,[23] since it is provoked by the fission axis, i.e. the axis along which the fission fragments are flying apart in opposite directions, to rotate in a plane perpendicular to the spin of the polarised neutron inducing fission. If the nucleus undergoing fission is rotating, the α-particles will experience a rotating Coulomb field and at least partly will follow the rotation. Trajectory calculations for α-particles moving in the rotating Coulomb field provided by the two main fragments confirm the interpretation. For the time being, precise measurements of the ROT effect have proven to be a reliable tool to inquire into specific details of the fission process, i.e. representing a valuable contribution to the spectroscopy of channel states in low energy fission. The values for the ROT shift obtained for the two reactions ^{235}U(n,f) and ^{239}Pu(n,f) are in good agreement with trajectory calculations for the relevant sets of channel spin (J,K) values.[24,25]

References

1. C. Wagemans, in The Nuclear Fission Process, edt. C. Wagemans, CRC Press, Boca Raton, Fl. USA, 1991, Chap. 12.
2. M. Mutterer, and J. Theobald, in Nuclear Decay Modes, edt. D.N. Poenaru, IOP, Bristol, UK, 1996, Chap. 12.
3. F. Gönnenwein, M. Mutterer, and Yu. Kopatch, Europhysics News **36/1**, 11 (2005).
4. C. Wagemans et al., Nucl. Phys. **A 742**, 291 (2004).
5. P. Heeg et al., Nucl. Instr. Meth. in Phys. Research **A 278**, 452 (1989).
6. M. Mutterer et al., Phys. Rev. **C 78**, 064616 (2008).
7. P. Singer et al., Proc. DANF96, Častá Papiernička, Slovakia, (JINR, Dubna, 1996), p. 262; P. Singer, Ph.D. Thesis, TU Darmstadt (1997).
8. Yu.N. Kopatch et al., Phys. Rev. Lett. **82**, 303 (1999).
9. Yu.N. Kopatch et al., Phys. Rev. **C 65**, 044614 (2002).
10. M. Mutterer et al., Nucl. Phys. **A 738**, 122 (2004).
11. E. Cheifetz et al., Phys. Rev. Lett. **29**, 805 (1972).
12. A.P. Graevskii, and G.E. Solyakin, Sov. J. Nucl. Phys. **18**, 369 (1974).
13. I. Halpern, Annu. Rev. Nucl. Sci. **21**, 245 (1971).
14. V.A. Rubchenya, and S.G. Yavshits, Z. Phys. **A 329**, 217 (1988).
15. A. Pik-Pichak, Phys. Atom. Nucl. **57**, 906 (1994).
16. G. Valskii, Sov. J. Nucl. Phys. **24**, 140 (1976).
17. H.H. Knitter, U. Brosa, and C. Budtz-Jrgensen, in The Nuclear Fission Process, edt. C. Wagemans, CRC Press, Boca Raton, FL., USA, 1991, Chap. 11.
18. Yu.N. Kopatch et al., Acta Physica Hungarica, New Series - Heavy Ion Physics **18**, 399 (2003).
19. Yu.N. Kopatch et al., AIP Conf. Proc. **798**, 115 (2005).
20. M. Speransky et al., Proc. ISINN12, Dubna, Russia, JINR Dubna 2004, p. 430.
21. P. Jesinger et al., Yad. Fiz. **65** 662 (2002); Phys. Atom. Nucl. **65**, 630 (2002).
22. P. Jesinger et al., Nucl. Instr. and Meth. in Physics Research **A 400**, 618 (2000).
23. F. Gönnenwein et al., Phys. Letters **B 652**, 13 (2007).
24. F. Gönnenwein et al., Proc. EXON09, Sochi, Russia, 2009, World Scientific, Singapore, in press.
25. A.Gagarski et al., AIP Conf. Proc. **1175**, 323 (2009).

Neutron Emission in Fission

# PROMPT NEUTRON EMISSION IN ^{252}CF SPONTANEOUS FISSION

F.-J. HAMBSCH, S. OBERSTEDT

EC-JRC-Institute for Reference Materials and Measurements, Retieseweg 111, B-2440 Geel, Belgium

SH. ZEYNALOV

JINR-Joint Institute for Nuclear Research, Dubna, 141980, Russia

The prompt neutron emission in spontaneous fission of ^{252}Cf has been investigated applying digital signal electronics. The goal was to compare the results from digital data acquisition and digital signal processing analysis with results of the pioneering work of Budtz-Jørgensen and Knitter. Using a twin Frisch-grid ionization chamber for fission fragment (FF) detection and a NE213-equivalent neutron detector in total about 10^7 fission fragment-neutron coincidences have been registered. Fission fragment kinetic energy, mass and angular distribution, neutron time-of-flight and pulse shape have been investigated using a 12 bit waveform digitizer. The signal waveforms have been analyzed using digital signal processing algorithms. The results are in very good agreement with literature. For the first time the dependence of the number of emitted neutrons as a function of total kinetic energy (TKE) of the fragments is in very good agreement with theoretical calculations in the range of TKE from 140-220 MeV.

1. Introduction

The digital signal processing (DSP) method, along with a twin Frisch-grid ionization chamber (TGIC), was used to measure the kinetic energy-, mass- and angular distributions of fission fragments in correlation with prompt fission neutrons emitted in spontaneous fission of ^{252}Cf. The experiment was basically adopted from Ref. [1] with the replacement of the traditional analogue signal processing instrumentation with digital signal processing hardware and associated software. The anode current caused by a fission fragment (FF) in the TGIC was amplified by a charge-sensitive pre-amplifier and sampled with a 12 bit, 100 Ms/sec waveform digitizer (WFD). The step-like long anode signals were software-wise transformed into short current pulses, which allowed effective pulse pile-up elimination. The anode current pulses were used to determine the fission fragment angle with respect to the cathode-plane normal.

The prompt fission neutron (PFN) time-of-flight (TOF) spectroscopy was done after passing the neutron detector (ND) pulse, digitized with a 12 bit, 100 Ms/sec WFD, through a 12th order digital low pass filter. The ND pulse shape discrimination was implemented using raw signal waveforms. The measurement of the FF characteristics, both in coincidence and non-coincidence with the PFN, was done without re-adjusting the apparatus. A Cf-sample, deposited on a 100 µg/cm2 thick Ni foil backing, with an activity of about 500 fission/s, was mounted on the common cathode of a twin Frisch-grid ionization chamber (TGIC), which operated under normal conditions with P-10 as working gas at a constant flow of ~ (50 - 100) ml/min. About 1.2×10^7 coincidences between FF and PFN signals were acquired in the measurement, which is comparable to the statistics mentioned in Ref. [1]. A scheme of the set-up is given in Fig. 1.

2. Data analysis and results

The FF kinetic energy and the angle between a FF and the TGIC axis was analyzed using DSP algorithms described in Ref. [2]. The waveforms $P_{1,2}[k]$ consisting of 1024 samples each, representing the step-like current integral, were converted into the current pulse $I_{1,2}[k]$ using digital differentiation. Then $I_1[k]$ and $I_2[k]$ were digitally integrated using a 4th order low pass RC4 filter with a time constant of RC ~ 2.0 µs to obtain the total charge released by the corresponding FF during deceleration in the TGIC.

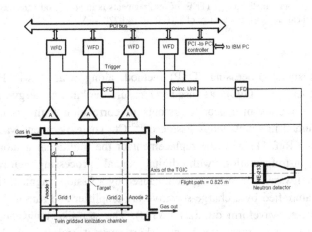

Figure 1. Simplified sketch of the experimental setup, which consists of the twin Frisch-grid ionization chamber (left) and a NE213-equivalent liquid scintillation detector (right)

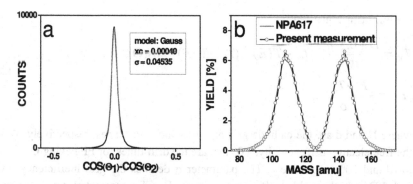

Figure 2. (a) Precision of fission-fragment angle measurement; (b) fission-fragment pre-neutron mass distribution

To measure the angle Θ_i between the FF and the TGIC's axis the drift time method, utilizing the drift time dependence on Θ_i of electrons released by the FF, was implemented:

$$T_i[L] = \sum_{k=Tg}^{Tg+L} k * I_i[k] \bigg/ \sum_{k=Tg}^{Tg+L} I_i[k] - T_g, \qquad (1)$$

where T_g is the trigger signal leading edge position. The trigger signal was obtained from the common cathode pulse and referred to as the time instant of the fission event. The dependence of $\cos(\Theta_i)$ on the drift time can be found using

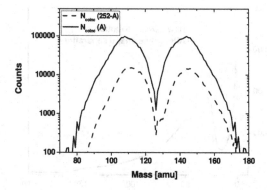

Figure 3. Mass distribution measured in coincidence with PFN for FF emitted toward the ND (solid line) in comparison with the mass distribution for FF emitted in the opposite direction (dashed line)

the following equations:

$$T_{90} = \frac{D + 0.5 \cdot d}{W}, \quad T_{90} - T(E) = (T_{90} - T_0(E)) \cdot \cos(\Theta),$$

$$P^C = \frac{P^O \cdot T_{90}}{T_{90} + \sigma \cdot T}, \quad (2)$$

where D and d are the cathode-grid and grid-anode distances, respectively. W is the free electron drift velocity. T0, T90 are the drift times for FF having Θi equal to 0° and 90°, respectively. The parameter σ denotes the grid inefficiency. PC and PO are the grid inefficiency corrected and uncorrected total charges collected on the anodes and, T is the drift time for the considered FF. The above equations were obtained taking the drift of electrons in electrostatic fields created between the cathode-grid and grid–anode spaces of the TGIC into account. Fig. 2a shows resulting distributions for the angular resolution in $\cos(\Theta)$. Fig. 2b shows the achieved pre-neutron mass distribution compared to the results of Ref. [3]. Finally, Fig. 3 shows the mass distribution stemming from FF in coincidence with a PFN (full line) and from those coming from the complementary fragment (dashed line) which needs to be corrected for.

The PFN kinetic energy was analysed from the TOF measurement using the prompt fission neutron distribution. The prompt fission γ-peak position and its width broadening dependence on prompt fission γ-ray pulse height were measured and parameterized. Then the PFN TOF distribution was considered as

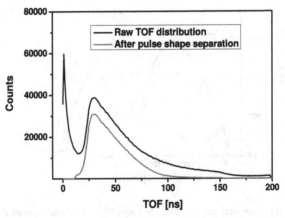

Figure 4. Prompt fission neutron TOF distribution illustrating the n-γ separation; the remaining background is negligible.

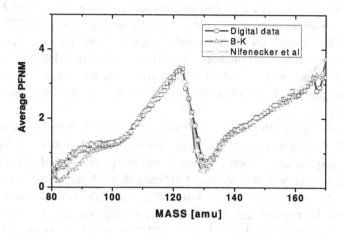

Figure 5. Prompt fission neutron multiplicity as a function of the fragment mass

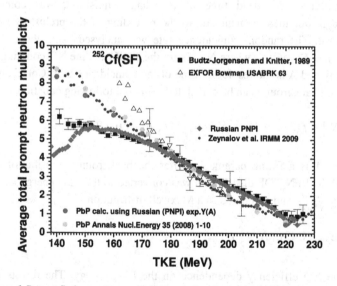

Figure 6. Prompt fission neutron multiplicity as a function of the total kinetic energy (TKE)

a distribution, satisfying the following integral equation:

$$f_s(t) = \int_0^\infty F(\tau)h(t-\tau)d\tau, \quad \text{with} \int_0^\infty h(t-\tau)d\tau = 1, \qquad (3)$$

where $f_s(t)$ is the measured TOF distribution after background subtraction. $F(\tau)$ is the original TOF distribution and h(t-τ, λ, η) is the resolution function (RF).

The RF depends on the total light output (λ) and on the flight path fluctuation due to the finite thickness of the neutron detector (ND) (η). For each FF in coincidence with the ND the following values were recorded: two FF kinetic energies FF_1, FF_2, PFN TOF value, total light output λ and pulse shape information.

The pulse shape information was used to separate PFN from prompt fission γ-rays and allowed to suppress the γ-radiation by about 240 times (see Fig. 4). The remaining background was considered as of two parts: The first part consists of random coincidences between the TGIC and either delayed neutrons from the target or the radioactivity of the surrounding materials, which form a time independent background and coincidences correlated with the fission events (when a fission event was detected by the TGIC, but a PFN hitting the ND after being scattered by air or other surrounding material). Uncorrelated coincidences, formed by a fission event trigger with a PFN detected from the next fission occurring during the dead time of the data acquisition, was completely eliminated in our measurement due to the recording of the prehistory of each fission event. The random coincidence rate was analysed using the following equation: $R = N_1 \cdot N_2 \cdot \Delta$, where N_1, N_2 are the TGIC and the ND counting rates, respectively and Δ is the resolution time of the coincidence unit. Correlated and uncorrelated background can be calculated using the following equation:

$$N_{un}(t) \approx N_1 \int_0^\Delta \tau(f(t-\tau)-R)d\tau, N_c \approx N_1 \int_0^\Delta \tau(f(\alpha(t-\tau))-R)d\tau, \quad (4)$$

where $\alpha > 1$ is a fitting parameter. After the background contributions were subtracted the PFN TOF distribution was converted to the PFN energy spectrum divided by $\sqrt{E_{PFN}}$ and compared to a Maxwell distribution,

$$N\exp(-E_{PFN}/1.42), \quad N = \int_0^\infty f(t)dt, \quad (5)$$

to find the ND efficiency dependence on the PFN energy. The results of the present measurement of the PFN multiplicity distributions are in very good agreement with those of Ref. [1]. The isotropic angular distribution of the PFN in the FF centre-of-mass reference system is confirmed.

The PFN multiplicity as a function of FF mass (see Fig. 5) is in reasonably good agreement with Ref. [1]. For the first time no reduction of the neutron multiplicity at low TKE was observed (see Fig. 6). This is expected from energy

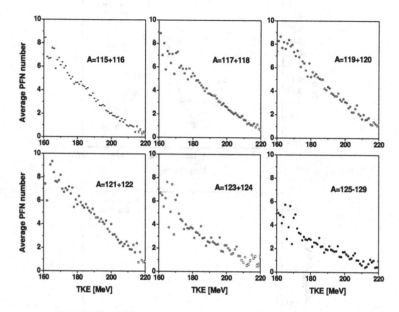

Fig. 7. PFN multiplicity as a function of TKE for selected masses.

balance considerations. A difference compared to Ref. [1] is also observed in the slope of the n(TKE). In the present work we deduce an energy cost per neutron of about 8.5 MeV in contrast to Ref. [1] where a value of about 13 MeV is reported. The latter is difficult to understand if one considers the binding energy of a neutron (~ 6 MeV) and the mean kinetic energy of the emitted neutron (~2.5 MeV). Our result is also backed up by recent calculations performed in Ref. [4]. From the obtained data a close to linear dependence of the PFN multiplicity on TKE was found for various mass ranges (see Fig. 7). That data was used to evaluate the neck elasticity as a function of FF mass (see Fig. 8a) and the maximum value of the TKE when neutron emission is still possible for fission fragments (see Fig. 8b). Due to the different slope the results given in Fig. 8a are also found to be different to the ones reported in Ref. [1].

3. Conclusions

The presented experiment shows a superiority of the digital signal processing method compared to the analogue one. A better neutron-γ separation is achieved compared to Ref. [1] and, for the first time, no reduction of the PFN multiplicity

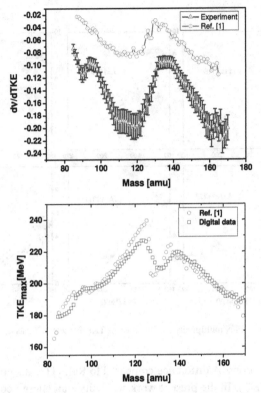

Figure 8. (a) Neck elasticity as a function of mass, (b) the maximum TKE, when FF stops PFN emission in comparison with literature

as a function of TKE is observed. The slope was found different to Ref. [1] but in good agreement with theoretical calculations.

Acknowledgments

Sh. Z. would like to thank the European Commission for granting him a Seconded National Expert contract.

References

1. C. Budtz-Jørgensen and H.-H. Knitter, *Nucl. Phys.*, **A490**, 307 (1988).
2. O.V. Zeynalova, Sh.S. Zeynalov, F.-J. Hambsch, S. Oberstedt, *Bulletin of Russian Academy of Science*: Physics **73**, 506 (2009).
3. F.-J. Hambsch, S. Oberstedt, Nucl. Phys., **A617**, 347 (1997).
4. A. Tudora, Ann. Nucl. Energy **35**, 1 (2008).

ENERGY MEASUREMENT OF PROMPT FISSION NEUTRONS IN ^{239}Pu(n,f) FOR INCIDENT NEUTRON ENERGIES FROM 1 TO 200 MEV

A. CHATILLON, T. GRANIER, J. TAIEB, G. BELIER, B. LAURENT

CEA,DAM,DIF,F -91297 Arpajon, France

S. NODA, R.C. HAIGHT, M. DEVLIN, R.O. NELSON, J.M. O'DONNELL

LANSCE, Los Alamos National Laboratory,MS H855, Los Alamos, NM 87545, USA

Prompt fission neutron spectra in the neutron-induced fission of ^{239}Pu have been measured for incident neutron energies from 1 to 200 MeV at the Los Alamos Neutron Science Center. Mean energies obtained from the spectra are discussed and compared to theoretical model calculation.

1. Introduction

An experimental campaign was started in 2002 in the framework of a collaboration between CEA-DAM and the Los Alamos National Laboratory to measure the prompt fission neutron spectra (PFNS) for incident neutron energies from 1 to 200 MeV with consistent error uncertainties over the whole energy range. The prompt neutron spectra in ^{235,238}U(n,f) and ^{237}Np(n,f) have been already studied successfully [1-4]. A first attempt to characterize the prompt neutrons emitted during the fission of the ^{239}Pu was done in 2007 [5-7]. This contribution will focus on the results obtained during the final experiment to measure the PFNS in ^{239}Pu(n,f) performed in 2008.

2. Experimental Setup

2.1. Production of neutron beam at the WNR facility

The WNR facility at LANSCE provides pulsed, white neutron beams with an energy distribution spread from 0.5 to several hundreds of MeV. The maximum intensity is around 2 MeV. Neutrons are produced thanks to spallation reactions induced by an 800 MeV pulsed proton beam on a tungsten target. They are collimated on several beam lines (see Figure 1). Our experiments were set on the

30° right flight path at 22.7 meters from the spallation target with a collimation resulting in a 2.8 cm diameter beam spot at the ^{239}Pu target position.

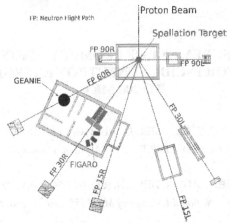

Figure 1. Schematic view of the WNR facility. FIGARO is located on the 30° right flight path.

2.2. Fission chamber

The target consists of an ionization fission chamber containing 100 mg of ^{239}Pu deposited onto multiple metal plates. The spacing between the plates is 1.7 mm. The body of the chamber is made of aluminum. The applied voltage between the electrodes was +300 V. The signal of the fission chamber was used to trigger the data acquisition system (based on the MIDAS software).

2.3. Prompt fission neutron detection with FIGARO

The neutron detector array FIGARO [7,8] was used to measure the prompt fission neutrons in coincidence with the fission chamber signals (see Figure 2).

Figure 2. Picture (left) and schematic view (right) of the experimental set-up.

FIGARO consists of twenty EJ301 organic liquid scintillators located about one meter from the fission chamber, on seven different detection angles (45°, 60°, 75°, 90°, 105°, 112° and 135° with respect to the beam direction).

In 2008, three EJ301 detectors were replaced by solid organic scintillators: one stilbene detector and two paraterphenyl detectors which are known for having high efficiency and good neutron-gamma discrimination at low energy (below 1 MeV). The comparison of the EJ301 and paraterphenyl detectors raw spectra, taken with a ^{252}Cf source, shows this feature (see Figure 3). Whereas at 1 MeV, the efficiency of the EJ301 starts already to decrease because of the higher threshold for neutron-gamma discrimination, this falling does not happen before 700 keV in the case of the paraterphenyl detector.

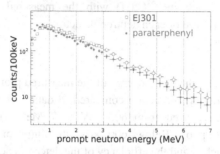

Figure 3. Comparison of raw energy spectra of prompt fission neutrons emitted in ^{252}Cf(sf). Open squares are the standard EJ301 scintillator detectors data and dots correspond to the paraterphenyl data. The threshold is equivalent to 300 keV neutron energy for both detectors.

These detectors are sensitive to neutrons and γ-rays but with different responses. Thanks to this feature, neutrons and γ-rays could be discriminated offline via a pulse shape analysis based on the charge integration of the short-time and long-time components of the pulses.

3. Analysis

For each triggered event, two times of flight were measured (with a resolution of 3.5 ns FWHM). The first one corresponds to the time-of-flight of the incoming neutron from the spallation target to the fission chamber. The second is the time-of-flight of the prompt fission neutron from the fission target to the scintillator. From these times of flight the velocities and kinetic energies of both incident and emitted neutrons were determined event by event.

A non-negligible background component due to the incident neutrons scattering on the fission chamber structure (windows and samples' backings) is visible. These scattered neutrons are not correlated with fission and create background in the neutron detectors by random coincidences. In order to measure this background an additional trigger was used. It consists of an electronic pulser sampling time uniformly. The data recorded from this trigger allowed us to monitor the background during the experiment and to determine its contribution as a function of the incident neutron energy. Spectra of the scattered neutrons were then subtracted offline from the spectra measured in coincidence with the fission chamber. The signal-to-background ratio was on the order of 2.

PFNS were determined for several incident neutron energy groups from 1 to 200 MeV. Afterwards the spectra were corrected for the detector efficiency using data taken with a spontaneous fission source of ^{252}Cf, placed at the position of the fission chamber. For these runs, a γ-ray detector located at 20 cm from the source was triggering the experiment. The comparison of the well known prompt fission neutron spectrum in ^{252}Cf(sf) with the measured spectra gives the efficiency correction curve of each neutron detector.

4. Results

Figures 4 and 5 present the mean energy of the measured prompt fission neutron spectra after background subtraction compared to data evaluations. The EJ301 neutron detectors were used from 600 keV to 7 MeV. Above 7 MeV the statistics are really poor. Below 600 keV the limit of the pulse shape discrimination is reached and the efficiency of the detectors decreases strongly.

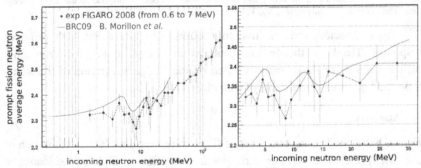

Figure 4. Mean energy of the prompt neutrons emitted in ^{239}Pu(n,f) calculated over the range 600 keV to 7 MeV. Experimental data (dots and dashed line) are compared with the BRC09 evaluation (solid line). Left panel: for the whole energy range. Right panel: for incident neutron energy below 30 MeV.

In Figure 4 the mean energy of the experimental spectra from 600 keV to 7 MeV is represented as a function of the incident neutron energy. The same energy cut is applied to the model calculations represented in the figures with the solid lines. The evaluated results are based on the Los Alamos model [9] in its improved form, following the prescription of A. Tudora and G. Vladuca [10], and implemented at the CEA/DAM of Bruyères-le-Châtel by B. Morillon.

In order to obtain an estimate of the mean neutron energy without any energy cut, the spectra have been fitted with a Maxwellian distribution:

$$N(E) = 2A\sqrt{\frac{E}{\pi T_m^3}} \exp(-\frac{E}{T_m}).$$

A is the integral of the function and T_m is the so-called "fission temperature" which is related to the emitter temperature, and is directly proportional to the total mean energy. Results are represented in Figure 5.

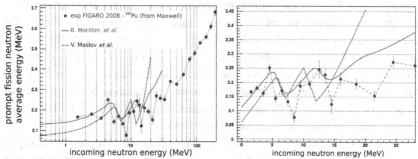

Figure 5. Mean energy of the prompt neutrons emitted in ^{239}Pu(n,f). Experimental data (dashed lines), deduced from the Maxwellian fit, are compared with evaluation of V. Maslov [11] (dotted lines) and B. Morillon (solid lines).

5. Discussion

Prompt fission neutrons are emitted from different sources. Mainly they are evaporated by the fully accelerated fission fragments, but depending on the incident neutron energy, they can also be emitted before fission. When the incident neutron energy is higher than the neutron separation energy, the compound nucleus gets enough excitation energy to evaporate one neutron and a competition starts between the first chance fission (n,f) and the second chance fission (n,nf) reactions. Then at higher neutron beam energy (already above 15 MeV), pre-equilibrium contribution starts to increase and is not anymore negligible compared to the two neutron evaporation contributions (fragments and compound nucleus).

Pre-equilibrium neutrons are largely emitted at forward angles with energies up to the incident neutron energy, whereas the neutron detection system is located at side angles and efficient from 500 keV up to 7 MeV. Our set-up is therefore mainly sensitive to the evaporation contributions, which are isotropic and follow (at first order) a Maxwellian distribution. As a consequence, the data presented here do not include all the contributions to prompt neutron spectrum and starting from ~15 MeV beam energy, the mean energies for the measured prompt fission neutrons are below the real ones, as pre-equilibrium neutrons are more energetic than the evaporated neutrons.

The right-hand panels of Figures 4 and 5 show that the mean energy is strongly dropping around 7 and 14 MeV. This behavior has already been seen experimentally for other actinides and was predicted by the models (see for example [1-4]). These two dips sign the opening of fission of second and third chances. Their depth show the weight of the pre-fission neutrons, since the neutrons evaporated by the compound nucleus are emitted with less kinetic energy mainly because the emitter temperature is cooler, but also because these neutrons are emitted in the laboratory frame. The agreement between the experimental values and the calculation is fair, especially with the BRC09 evaluation (better than 4%) that predicts the position of the openings of second and third chances more accurately than the calculations of V. Maslov [11]. But the amplitude of the third chance is better reproduced by the model of V. Maslov, whereas BRC09 seems to underestimate it. Above 15 MeV, mean kinetic energy seems overestimated by both models.

The general trend is an increase of about 30 % in the mean kinetic energy of the prompt fission neutrons over the whole incident neutron energy range (see left panels of Figures 4 and 5). On average, the temperature of the compound nucleus is increasing with the beam energy, so the prompt neutrons will globally be emitted with higher kinetic energy for higher incident neutron energies. But the rise of the mean energy seems to start only around 20-25 MeV. Below this limit and except for the two dips around 7 and 14 MeV, the mean energy of the prompt neutron is rather flat around 2.15 MeV. Such a behavior is in agreement with the energy sorting hypothesis from B. Jurado and K.-H. Schmidt [12-13]. However the data uncertainties are too large to conclude unambiguously.

6. Conclusion

The kinetic energy spectra of the prompt neutrons emitted during the neutron-induced fission of the ^{239}Pu actinide were measured depending on the beam energy with the FIGARO set-up at LANSCE.

The prompt fission neutron average energy was deduced from the spectra for a large range of incident neutrons energies. The positions of the openings of the fission of second and third chances were measured at 7 and 14 MeV, respectively. These experimental values are well reproduced (better than 4 %) by the BRC09 evaluation. However, except at the two dips at 7 and 14 MeV, the experimental average energy seems rather flat up to 20 MeV whereas evaluations predict an increase of the mean energy. The uncertainties are unfortunately too large to conclude.

We plan to extend the comparison to model to the PFNS shape. Spectra are indeed potentially more sensitive observables than mean neutron energy.

Acknowledgments

This work was performed under the auspices of a cooperation agreement between CEA/DAM and DOE/NNSA on fundamental sciences and benefitted from the use of the LANSCE accelerator facility and was performed in part under the auspices of the US Department of Energy at the LANL.

References

1. Th. Ethvignot *et al.*, Phys. Lett. B **575**, 221 (2003).
2. Th. Ethvignot *et al.*, Phys. Rev. Lett. **94**, 052701 (2005).
3. Th. Ethvignot *et al.*, Phys. Rev. Lett. **101**, 039202 (2008).
4. J. Taieb *et al*, in *Proceeding of the International Conference in Nuclear Data for Science and Technology, Nice, France, 2007*, edited by O. Bersillon, F. Gunsing, E. Bauge, R. Jacqmin, and S. Leray, (EDP Sciences), pp 429-432.
5. R. C. Haight, S. Noda, and J. M. O'Donnell, "Los Alamos Analysis of an Experiment to Measure Fission Neutron Output Spectra from Neutron-Induced Fission of ^{235}U and ^{239}Pu," Los Alamos National Laboratory report LA-UR-08-2585, April 16, 2008.
6. A. Chatillon *et al.*, in *Proceeding of the 12th International Conference On Nuclear Reaction Mechanism, Varenna, Italy, 2009*, edited by F. Cerutti and A. Ferrari.
7. S. Noda, R. C. Haight *et al.*, LA-UR-10-03762 (submitted for publication, 2010).
8. D. Rochman *et al.*, Nucl. Instrum. Meth. Phys. Res., Sect A **523**, 102 (2004)
9. D.G. Madland *et al.*, Nucl. Sci. Eng. **81**, 213 (1982).
10. A. Tudora *et al.*, Nucl. Phys. **A740**, 33 (2004).
11. V. Maslov *et al.*, Phys. Rev. C **69**, 034607 (2004)
12. K.-H. Schmidt and B. Jurado, Phys. Rev. Lett. **104**, 212501 (2010).
13. B. Jurado and K.-H. Schmidt (these proceedings).

ANISOTROPIC NEUTRON EVAPORATION FROM SPINNING FISSION FRAGMENTS

L. STUTTGÉ, O. DORVAUX
Département de Recherches Subatomiques, IPHC-Université de Strasbourg, 67037 Strasbourg Cedex 2, France

F. GÖNNENWEIN
Physikalisches Institut, Universität Tübingen, 72076 Tübingen, Germany

M. MUTTERER
Institut für Kernphysik, Technische Universität, 64289 Darmstadt, Germany
and
GSI Helmhotzzentrum für Schwerioneforschung, 64291 Darmsatdt, Germany

YU. KOPATCH
Frank Laboratory of Neutron Physics, JINR, 141980 Dubna, Russia

E. CHERNYSHEVA
Flerov Laboratory of Nuclear Reactions, JINR, 141980 Dubna, Russia

F. HANAPPE
PNTPM, Université Libre de Bruxelles, 1050 Brussels, Belgium

F.-J. HAMBSCH
EC-JRC-Institute for Reference Materials and Measurements, 2440 Geel, Belgium

Neutron evaporation anisotropy in the centre of mass of the rotating fission fragments in the spontaneous fission of ^{252}Cf has been investigated within the CORA experiments. If it is well accepted that the bulk of emitted neutrons originate from an isotropic evaporation in the centre of mass of the moving fragments, discrepancies in experimental as well as in theoretical energy and angular distributions appear throughout many attempts performed by various authors. Scission neutrons most probably contribute but don't allow to explain totally the observed anisotropy. Due to its weak contribution to the total anisotropy, the centre of mass anisotropy is very difficult to be highlighted. A novel experimental approach has been developed to extract this effect and will be presented as well as some first results.

1. Neutron emission in the fission process

During the fission process, neutrons are mainly evaporated by the fully accelerated fission fragments. This emission is isotropic in the centre of mass of the moving fragments. The transformation of the kinematics into the laboratory system leads to an anisotropic angular distribution, the kinematic anisotropy.

This has already been shown in 1962 by H.R. Bowman et al [1] for the ^{252}Cf spontaneous fission. But already in these measurements deviations from a pure isotropic evaporation appear. To bring out the nature of these deviations H.R. Bowman et al performed calculations in which they introduced a scission neutron contribution. Indeed neutrons are probably also ejected at an earlier stage of the fission process at the scission point. A contribution of 10% scission neutron emission explained the excess of neutrons observed around 90°. But the remaining deviations at small angles around the heavy and light fragments couldn't be removed. Although these features are in favor of an anisotropy in the centre of mass of the fission fragments, the experimental uncertainties didn't allow to dare any further conjecture on the nature of this deviation.

Many experimental results have been accumulated since H.R. Bowman and collaborator's work leading to contradictory conclusions. Measurements on the ^{252}Cf spontaneous fission have been performed by C. Budtz-Jørgensen and H.-H. Knitter [2] in 1988. They came to the conclusion that the only transformation from the laboratory system to the centre of mass system (cms) of the fission fragments resulted in a pure isotropic emission in the centre of mass of the fission fragments.

J. Terrell [3] tried to introduce an anisotropy in the centre of mass of the fission fragments for different systems. He found that the effect of anisotropy was most important at low neutron energies. But he was not very confident on the possibility to extract any information on anisotropy from the neutrons spectra alone, as the uncertainties and approximations of the evaporation theory were too important.

K. Skarsvag et al [4] found that introducing a 15% scission neutron contribution to the isotropic evaporation by the fission fragments in the neutron induced fission of ^{235}U allowed a perfect agreement between the fit and the experimental spectrum as shown in the left spectrum of Figure 1.

In recent calculations V. Bunakov et al. [5] produced theoretical angular distributions for ^{235}U which are presented on the left part of Figure 1. It appears that the centre of mass anisotropy will reinforce the kinematic anisotropy in the laboratory system but the effect is very weak.

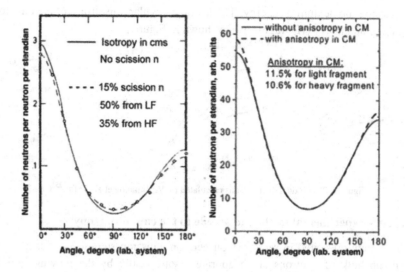

Figure 1. Number of neutrons versus angle in the neutron induced fission of ^{235}U. – left:. The points are experimental. The solid curve corresponds to isotropic evaporation from fragments and the dashed curve to emission during the fission process and isotropic evaporation. From reference [4]. – right: Calculated number of neutrons without anisotropy in the c.m. (solid curve) and with anisotropy in the cms (dashed curve). From reference [5].

2. Centre of mass anisotropy of neutron evaporation

It is well established that fission fragments carry high angular momenta (6-8$\hbar$). Neutrons evaporated from a rotating nucleus will preferentially be emitted in the equatorial plane of spin as represented on the left part of Figure 3. For a fixed spin this centre of mass anisotropy can be expressed as $1+A.\sin^2\theta_{cm}$, where θ_{cm} is the polar angle relative to the spin. The averaging on all possible orientations of the spin perpendicular to the fission axis will produce a forward-backward increase along the fission axis in the cms and the averaging over all spins will lead to an anisotropy expressed as $1+b.\cos^2\theta_{cm}$, where b is the anisotropy parameter. This means that even in the cms of the fission fragments the evaporation will be anisotropic.

These considerations have been applied by V. Bunakov et al [5], who calculated the anisotropy as a function of angle for different angular momenta and neutron energies as shown in Figure 2. For l=0 there is no anisotropy. With increasing momentum (Figure 2-a)) or energy (Figure 2-b)), the anisotropy increases. Unfortunately the number of neutrons decreases quickly with increasing momentum or energy. The average angular anisotropy in the cms for light (L) and heavy (H) fragments are shown in Figure 2-c).

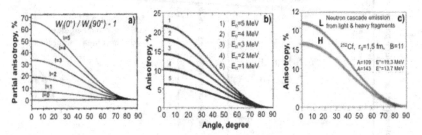

Figure 2. Theoretical distributions calculated by V. Bunakov et al [5] for ^{252}Cf.

3. New experimental method to accede to the cms anisotropy

The review presented in section 1 on neutron emission shows that if it is clear that the bulk of neutrons are evaporated isotropically by the moving fission fragments the existence of a scission point emission is less clear. Even less evident is the existence of any anisotropy in the cms of the fission fragments. One has to have in mind that if this dynamic anisotropy exists it is extremely weak and difficult experimentally to disentangle from kinematic anisotropy as shown in Figure 1. Theoretical calculations contain uncertainties and approximations which don't allow any definite conclusion. The many attempts to put in high light dynamic anisotropy were based on angular and energy spectra of the emitted neutrons.

To get rid of the kinematic anisotropy contribution a new experimental method has been developed. It is based on the analysis of triple coincidences between one fission fragment and two fission neutrons. Figure 3 explains the principle of the method. If one considers a given spin and an extreme cms anisotropy, all neutrons will be emitted in a plane perpendicular to the spin. If one projects the fission axis and all neutron events on a plane perpendicular to the fission axis, all neutron events will be aligned on a single line, the x-axis in Figure 3. To any spin orientation corresponds a plane of neutron emission and in the projection a single line of neutron emission perpendicular to the spin. The projection lines of neutron emission follow the inclination of the fragment spin. Figure 3 shows also the behavior in case of perfect isotropy where the neutrons are spread isotropically all over the x-y plane. If triple coincidences between one fission fragment and two fission neutrons are selected, the difference of the azimuthal angle of the two neutrons, $\Phi_{21}=\Phi_2-\Phi_1$, will follow the same distribution whatever the spin orientation: maxima at 0° and 360° when the neutrons are emitted on the same side of the y-axis, 180° when they are emitted on both sides. This distribution corresponds to extreme anisotropy and has to be compared to the flat distribution in the case of perfect isotropy. Of course in reality, the distribution will be a mixing of both situations and will be washed out as shown on the bottom of Figure 3. The success of the method will depend on which curve will be the real one: the very optimistic case of the solid curve, the worse case of the dotted curve in which case the method is hopeless or the in-between case of the dashed curve.

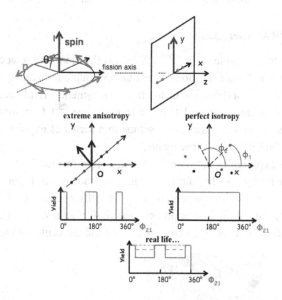

Figure 3. Extreme anisotropy in the cms: all neutrons are emitted in a plane perpendicular to the spin axis (top). Projection plane x-y : the projection lines follow the inclination of the spin orientations compared to the random distribution in the perfect isotropy case. Azimuthal angle difference distributions, Φ_{21}, in both extreme cases (centre) and real distribution (bottom).

Simulations have been performed to get an estimation of the effect [6]. Figure 4 shows the resulting projection of the neutron-neutron angular distribution onto the x-y plane starting with an original distribution $W \propto 1+A.\sin^2\theta$, with A=1 and where θ is the polar angle between the neutron and the spin. The projection distribution ends up as a sinusoidal distribution as $W \propto 1+a.\cos^2\varphi_{nn}$, with a=0.08 which is an order of magnitude smaller than A.

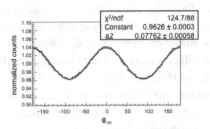

Figure 4. Simulated projection of relative neutron-neutron angular distribution onto the x-y plane.

4. CORA experiments

The neutron multidetector DEMON [7] is the ideal tool to accede to the neutron-neutron angular distributions. The CODIS double ionization chamber [8] offers a very accurate 4π detection of the fission fragments. Triple coincidences between one fission fragment and two neutrons in the ^{252}Cf spontaneous fission were recorded associating these two setups in a series of experiments dubbed CORA (CORrelation Array) experiments.

An exploratory experiment, CORA-1, performed in 2003 allowed to check the feasibility of the measurement. As shown on Figure 5, DEMON was mounted in a wall configuration of 74 cells and a set of 8 detectors on the opposite side. In a first approach one fission axis was considered by selecting fission fragments within a cone of 15° around the centre of the wall direction.

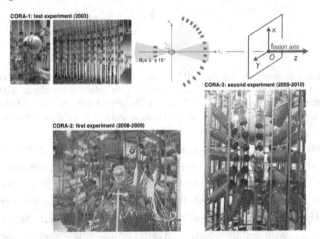

Figure 5. CORA experiments associating the neutron multidetector DEMON and the double ionization chamber CODIS.

Figure 6 shows the resulting neutron-neutron angular distribution (right spectrum) which has to be compared to the simulated distribution taking into account the experimental filter (left spectrum). The simulated distribution presents bumps around 0°, 180° and 360° where the anisotropy effect is expected, but also at 90° and 270°. This distribution should be flat; the fake bumps show that the efficiency correction was not perfect. This is due to the fact that the DEMON configuration was not adapted to the goal of the experiment as the geometry was to regular with many missing angles. The experimental distribution doesn't show any bump at 90° and 270° but maxima can be observed around 0°, 180° and 360° as is confirmed by the sinusoidal fit plotted on the experimental spectrum.

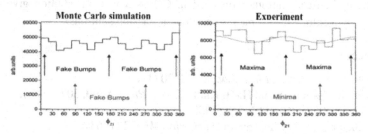

Figure 6. Neutron-neutron angular distributions resulting from a Monte-Carlo simulation (left) and from CORA-1 experiment.

This very promising result led to a dedicated experiment, CORA-2, with an optimized geometry. As shown on Figure 5, the DEMON detector was mounted in a forward and a backward wall configuration, in a more isotropic arrangement of the individual cells. Thus a maximum of different angles were accessible.

The bad point of such a compact configuration is the cross talk between two adjacent neutron detectors. This will be the main source of systematical uncertainty and has to be considered very carefully as it may contaminate the results on the very weak effect of the cms anisotropy. New dedicated measurements will be performed to quantify precisely the cross talk influence.

To optimize the set up, as appeared in new simulations, as well as to overcome any experimental bias another configuration DEMON has been used in CORA-3. The cylindrical configuration allows to increase the angular range of the fission fragments in coincidence with neutrons.

5. Summary and outlook

The novel experimental approach developed within the CORA collaboration and based on the measurement of triple coincidences between a fission fragment and to fission neutrons detected with the multidector DEMON associated to the fission spectrometer CODIS presents a very promising way to extract a possible cms anisotropy of the neutron evaporated by the moving fission fragments.

During the two dedicated experiments, CORA-2 and CORA-3, which lasted about 10 and 5 months respectively, about 10^9 triple coincidences have been recorded compared to 10^6 in CORA-1. The analysis of this huge amount of data has to be handled very carefully and are awaiting the cross talk measurements in order to produce reliable results.

The cylindrical configuration used in CORA-3 may in addition give the opportunity o extract also some information on scission neutrons.

References

1. H. R. Bowman et al, *Phys. Rev.* **126**, 2120 (1962).
2. C. Budtz-Jørgensen and H.-H. Knitter, *Nucl. Phys.* **A490**, 307 (1988).
3. J. Terrel, *Phys. Rev.* **113**, 527 (1959).
4. K. Skarsvag and K. Bergheim, *Nucl. Phys.* **45**, 72 (1963).
5. V. Bunakov et al, Proc. "Int. Sem. ISINN-13", Dubna, Russia,2005, p 173.
6. Yu. Kopatch et al, private communication.
7. I. Tilquin et al, *NIM* **A365**, 466 (1995).
8. Yu.N. Kopatch et al, *Phys. Rev.* **C65**, 044614 (2002).

New Facilities and Detection Systems

STATUS OF MYRRHA AND ISOL@MYRRHA IN MARCH 2010

H. AÏT ABDERRAHIM, J. HEYSE, J. WAGEMANS

Institute for Advanced Nuclear Systems, SCK•CEN, Boeretang 200, B-2400 Mol, Belgium

From 1998 on, SCK•CEN has been studying the coupling of a proton accelerator, a liquid Lead-Bismuth Eutectic (LBE) spallation target and a LBE cooled, sub-critical fast reactor. The project, called MYRRHA, aims at constructing an Accelerator Driven System (ADS) at the SCK•CEN site in Mol (Belgium). MYRRHA is designed as a multi-purpose irradiation facility in order to support research programmes on fission and fusion reactor structural materials and nuclear fuel development. Applications of these are found in ADS systems and in present generation as well as next generation critical reactors. The first objective of MYRRHA however will be to demonstrate on one hand the ADS concept at a reasonable power level and on the other hand the technological feasibility of transmutation of Minor Actinides (MA) and Long-Lived Fission products (LLFP) arising from the reprocessing of used fuel. MYRRHA will also help the development of the Pb alloys technology needed for the lead fast reactor Generation IV concept. Moreover, the implementation of this project will trigger the development of various applications, like a facility for fundamental physics at high-intensity Radioactive Ion Beams (RIB) called ISOL@MYRRHA. This paper presents the present status of the MYRRHA project and in particular the characteristics of ISOL@MYRRHA.

1. Introduction

One of the major challenges that our society faces is the increasing demand for energy in general and electricity in particular. During the last century our energy supply was based on fossil fuels. Nowadays, we are confronted with decreasing hydrocarbon reserves and excessive CO_2 emissions. At the same time renewable energy sources cannot satisfy the complete demand. For this reason the European Union, Japan, the United States, Korea, Russia, China, India and other countries recognize that nuclear energy needs to be part of the "energy mix" of the future.

The fact that most present nuclear power reactors only use a very limited fraction of natural uranium (basically ^{235}U that is present at 0.7% in natural uranium), cannot be ignored. This is the case for all present-day operational reactors with a thermal neutron spectrum. If nothing changes from a technological point of view and in the case of a rapid development of nuclear energy, known uranium resources will become scarce before the end of the century. Reactors with a fast neutron spectrum allow using a large part of the remaining 99.3% (^{238}U) of mineral uranium as fuel by transforming ^{238}U in ^{239}Pu. This technology allows using the present uranium resources up to 50 times more efficiently leading to uranium resources for more than several thousands of years. Moreover these reactors can work with thorium, which is four times more abundant on earth than uranium.

Apart from the large quantity of electricity, present nuclear reactors also produce high-level radioactive waste, for which a technical and socially acceptable solution is necessary. Even though the removal principles which are being considered at the moment, like disposal in deep geological layers, might offer a technically valid solution, the time scale needed for the radiotoxicity of the waste to drop to the level of natural uranium is very long (i.e. of the order of 500.000 to 1 million years).

Transmutation of high-level radioactive elements with a long half-life present in the used fuel (like the minor actinides americium, curium and neptunium) allows to significantly reduce this time scale. It changes from a 'geological' value to a value which is comparable to that of human activities. During transmutation the nuclei of these actinides are split in shorter lived fission products. The radiotoxicity of these is equal to that of the original uranium ore after few hundreds of years. In order to transmute these elements with a long half-life in an efficient way, nuclear reactors with a fast neutron spectrum are necessary. Moreover, the introduction of minor actinides in the reactor core requires appropriate measures to guarantee the flexible and safe operation of the system.

In order to develop the technologies associated with these fast reactors, an irradiation facility is needed. If one aims at transmuting large amounts of minor actinides in one installation, dedicated fast systems such as ADS systems need to be used. The design of such an installation for transmutation on a realistic scale is an essential step in the process that leads to the industrial application of this technique. It is not sufficient to demonstrate the principle of transmutation; also the necessary technology needs to be developed.

Apart from energy production, other high-end technologies require irradiation facilities, e.g. the development of astronautics and telecommunication materials or the development of radioisotopes for medical applications.

Each of these technologies needs the availability of an irradiation facility with enhanced performance in order to create irradiation conditions which are representative of the real situation when it comes to neutron energy and intensity. Therefore the design of a multipurpose experimental irradiation facility is an important step in the direction of the development of new technologies, opening the path to future research. This irradiation facility needs to have the largest possible flexibility and must be inherently safe.

In order to safeguard the irradiation capabilities for the future in Europe, Europe wants to realise a European Research Area on Experimental Reactors (ERAER). The flexible irradiation capacity within the ERAER is based on three pillars:

- A flexible thermal spectrum irradiation facility that will be answering the needs for industrial applications for present day (GEN II & III) reactors in terms of structural material and fuel performance improvement as well as some generic research for future (GEN IV) reactor concepts. It will also be acting as backup irradiation facility for radioisotopes production.
- A flexible fast spectrum irradiation facility to address the key issues related to material and fuel developments for GEN IV and fusion reactors and in a back-up role for radioisotopes production. Operation as an ADS system allows responding to the need expressed in the international community for an ADS demo at reasonable power level, demonstrating the ADS concept and the efficient transmutation of high-level nuclear waste (minor actinides).
- A dedicated irradiation facility for securing the radioisotopes production for medical applications in Europe and as a complementary facility in support of the industrial needs for technological development for present and future reactors.

The Jules Horowitz reactor in Cadarache, France, that started its construction and is expected to be in full operation by 2014/2015, is clearly able to respond to the need of a flexible thermal spectrum irradiation facility. The Netherlands have the intention to replace the HFR by the PALLAS reactor to secure the base-load radioisotope production in the future and hence clearly respond to the third pillar. In this context, it is clear that the MYRRHA project designed as a flexible fast spectrum irradiation facility will be able to fill in the niche described by the second pillar.

In March 2010, the Belgian federal government has committed itself to financing 40% of the total investment for MYRRHA. The remainder needs to be financed by an international consortium that has to be set up in the coming years. MYRRHA is foreseen to be fully operational by 2023.

2. MYRRHA, a safe and flexible research facility

MYRRHA will be a research facility that meets the requirements of a new irradiation facility mentioned above. In collaboration with different national and international partners, SCK•CEN, the Belgian Nuclear Research Centre in Mol has been working towards the realization of such a facility since 1998, by means of intense design activities and a significant R&D support programme. MYRRHA is based on the principle of an Accelerator Driven System, also called a 'hybrid reactor'.

2.1. The ADS principle

An ADS is a nuclear reactor with a subcritical core coupled to an external neutron source. In this context the term 'subcritical' means that, on average, for each generation of neutrons, less than one secondary neutron is capable of initiating a subsequent nuclear fission thus giving birth to a neutron of the next generation. In this way the chain reaction is not self sustaining, contrary to classical reactors where criticality allows for a generation of neutrons to provide on average one neutron to the next generation. In order for an ADS to operate continuously, an external neutron source is necessary. These supplementary neutrons are produced by spallation reactions, during which high energy protons, coming from a particle accelerator, are impinging on a heavy metal like lead. In the course of the spallation process the liquid metal nuclei emit a large number of neutrons whose energy spectrum is made of two parts: a rather conventional fission spectrum and a high energy tail up to the incident energy of the proton.

Since the functioning of an ADS is not dependent on the conservation of criticality, the machine can be exploited in a safe and flexible way under all circumstances and the system remains controllable, even when loaded with a substantial amount of minor actinides having a smaller delayed neutron fraction (30 to 100 pcm) than uranium (~700 pcm). In short, an ADS is an ideal candidate for the transmutation of highly radioactive waste.

2.2. Technology

MYRRHA consists of a proton accelerator with a proton energy of 600 MeV and an intensity of 2.5 mA, coupled to a liquid Lead-Bismuth Eutectic (LBE) spallation source. The spallation target is located in the centre of a subcritical reactor core with a fast neutron spectrum and cooled with liquid lead-bismuth. The reactor is a pool type reactor; the lead-bismuth of the spallation source and the one cooling the core flow in separate circuits. The concept is illustrated in Figure 1.

At the present status of the design, the reactor core consists of mixed oxide (MOX) fuel pins, typical of fast reactors, with a plutonium content of about 35 % and a length of 0.6 m. Apart from the spallation target the reactor vessel contains the reactor core which possesses several fast and thermal neutron irradiation positions. Also located in the vessel are the primary pumps and the heat exchangers for the cooling of the primary circuit of the reactor, as well as the robots for the handling of the fuel assemblies. The vessel has an inner diameter of 4.4 m and height of about 7 m and is installed in an underground excavation. It has a double wall and is surrounded by biological shielding in order to limit the surrounding dose as much as possible.

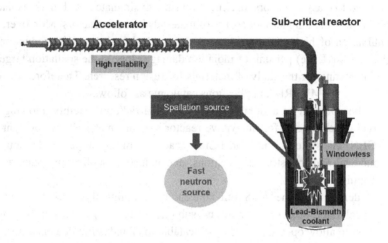

Figure 1. The MYRRHA accelerator driven system.

The full ADS R&D support programme is being carried out in collaboration with several European partners. At SCK•CEN the R&D programme in support of the MYRRHA project and ADS development in general, is focused on the following key issues:
- the design of the spallation target which is a typical part of an ADS that forms the link between classic reactor and accelerator technology;
- the development of lead-bismuth technology and conditioning;
- the behaviour of materials and lead-bismuth under irradiation;
- the qualification of fuel under irradiation in presence of lead-bismuth;
- the development of advanced instrumentation for ultrasound visualization and oxygen concentration monitoring in LBE;
- the development of robotics in aggressive environments (liquid metal and radiation);

- the investigation of core operation in subcritical and critical mode and the development of core monitoring techniques.

3. Research opportunities at MYRRHA

MYRRHA possesses several irradiation stations in and around the reactor core. In this way a very broad neutron spectrum, ranging from thermal energies to fast neutrons is available. Both thermal and fast neutron fluxes are very high compared to classic reactors, making it feasible to simulate – within a reasonable time frame – the long term exposure of materials in these reactors. Moreover, the combination of high radiation damage doses (dpa) and the high production of light gasses (H, He) per unit of radiation damage close to the spallation target is very interesting for the study of materials for fusion research. Therefore, one can summarize the MYRRHA applications catalogue as follows:
- to be operated as a flexible fast spectrum irradiation facility allowing for fuel developments for innovative reactor systems, material development for GEN IV systems and for fusion reactors, radioisotope production for medical and industrial applications, and industrial applications such as Si-doping;
- to demonstrate the ADS full concept by coupling the three components (accelerator, spallation target and sub-critical reactor) at reasonable power level to allow operation feedback, scalable to an industrial demonstrator;
- to allow study of the efficient transmutation of high-level nuclear waste, in particular minor actinides that would request high fast flux intensity ($\Phi_{>0.75MeV} = 10^{15}$ cm^{-2}s^{-1}).

4. ISOL@MYRRHA: an exotic island

Apart from the experimental and irradiation possibilities in the subcritical reactor, the MYRRHA proton accelerator on its own can be used as a supply of proton beams for a number of – nuclear physics – experiments. In order to explore new research opportunities offered by the accelerator, a pre-study was initiated within the framework of the "Belgian Research Initiative on eXotic nuclei" (BriX) network. This study investigates unique possibilities for fundamental research using high intensity proton beams. As a first step a dedicated workshop entitled "Nuclear Physics Research at the MYRRHA Accelerator" (SCK•CEN, Mol, Belgium, April 6-9, 2008 [1]) was organized where international specialists on radioactive ion beam (RIB) and neutron research were invited to present their views. The conclusion was that an interesting approach for fundamental research using the 600 MeV proton accelerator is the installation of an Isotope Separator On-Line (ISOL) system to produce intense low-energy Radioactive Ion Beams (RIB) available for experiments requiring very long beam times. This approach would complement other European RIB initiatives. The aim of the present study is to lay down the general lines for RIB production and for a RIB research program that will guide a follow-up feasibility study.

Because of the strong similarities of the driver accelerator parameters, the so-called ISOL@MYRRHA will follow closely the RIB production schemes that are developed and successfully used at the ISOLDE-CERN and TRIUMF facilities. It will be equipped with ruggedized target-ion source systems that allow the use of a selection of target materials, including actinide targets that can withstand the proton beam power. Two types of ion sources are foreseen: hot cavity surface ion sources directly coupled to the high temperature target container that would allow at the same time the implementation of resonant laser ion ionization (RILIS [2]) and a simple low-charge state electron resonance ion source (ECRIS [3]) coupled to the target container with a cold transfer line for gaseous beam production. By using a part (between 100-200 µA) of the 600 MeV proton beam ISOL@MYRRHA will produce a wide spectrum of intense and pure radioactive Q=1+ ion beams at 50 keV energies. The rationale behind the limited choice of simplified and ruggedized target-ion source systems for ISOL@MYRRHA is that the facility should deliver RIB for experiments needing very long beam times up to a few months. In order to make effective use of the precious beam time, the parallel multi-users aspect of ISOL@MYRRHA will be a key issue in the feasibility study.

Long beam times will be available for high-precision experiments, experiments hunting for extremely weak signals or experiments with an inherently low efficiency. In this sense ISOL@MYRRHA will be complementary to other existing or planned facilities. Different physics cases are currently under investigation of which only a few are listed here. Detailed decay spectroscopy of light nuclei to determine important but extremely weak decay branches or for precise gamma-ray energy measurements using detectors with ultimate energy resolution but limited efficiency like e.g. bent-crystal spectrometers. These experiments can eventually be performed in ion traps to avoid disturbing effects from the host lattice. The latter possibility is also important for specific aspects in weak interaction studies through precise nuclear beta decay measurements. Atomic spectroscopy of stable and radioactive isotopes in atom traps where the variety of ground or isomeric state spin values can be exploited is another interesting topic in the field of atomic physics, while implantation and detection of specific radioactive probes in different host materials is a topic of great interest to solid state physics and applications in nanotechnology. The above list is far from exhaustive and will certainly evolve in the course of the preparatory study.

Soon after the preparatory study is completed, a feasibility study will be launched including a large consultation of international experts and interested parties.

5. Summary

SCK•CEN is designing an ADS system called MYRRHA. It is made of a subcritical fast neutron reactor loaded with MOX fuel and driven by an external neutron source. These primary neutrons are produced by means of an intense 600 MeV proton beam bombarding a lead-bismuth target. The objective of the project is to create an international irradiation research facility in a European Research Area on Experimental Reactors.

The Belgian government has decided in March 2010 to fund this at a level of 40% of the total investment of 960 M€ (€-2009). The next stages of the project for the period 2010-2014 are:

- Completing the front-end engineering design (FEED) to allow the preparation of the components specification and contracts tendering by 2015;
- Securing the licensing of the facility by the end of this phase;
- Consolidating the international consortium up to a total investment cost coverage of 80%;
- Completing this first stage successfully will open the door for the construction of the facility starting in 2015.

Acknowledgments

The ISOL@MYRRHA feasibility study was performed in the framework of the Interuniversity Attraction Poles project financed by the Belgian Science Policy Office of the Belgian State to whom the authors are grateful.

References

1. http://iks32.fys.kuleuven.be/wiki/brix/index.php/Workshops.
2. V.N. Fedosseev et al., *Nuclear Instruments and Methods* **B266** (2008) 4378.
3. N. Lecesne et al., *Nuclear Instruments and Methods* **B266** (2008) 4338.

NEXT GENERATION FISSION EXPERIMENTS AT GSI: SHORT AND LONG TERM PERSPECTIVES

A. BAIL*, J. TAIEB, A. CHATILLON, B. LAURENT, G. BELIER

CEA/DAM/DIF, 91297 Arpajon, France
* E-mail: adeline.bail@cea.fr

L. TASSAN-GOT, L. AUDOUIN

IPNO, 5 rue Georges Clemenceau, F-91406 Orsay, France

B. JURADO, K.-H. SCHMIDT

CENBG, CNRS/IN2P3, Chemin du Solarium B.P. 120, 33175 Gradignan, France

J. BENLLIURE

Universidade de Santiago de Compostela, Spain

F. REYMUND

GANIL, CEA-DSM/CNRS-IN2P3, F-14076 Caen, France

D. DORE, S. PANEBIANCO

CEA Saclay, CEA-DSM/IRFU/SPhN, F-91191 Gif-Sur-Yvette, France

SOFIA, Studies On FIssion with Aladin, as well as FELISe, Fission@ELISe are both part of the forthcoming GSI fission experimental program. They will benefit from relativistic actinide beams available at GSI to induce electromagnetic fission in reverse kinematics. SOFIA will take place in Cave C in the curent GSI facility, while FELISe will be located at FAIR, the GSI part to come. Both will enable to determine nuclear charge and mass as well as kinetic energy for each fission fragment. The neutron multiplicity will be measured too, and the determination of the excitation energy of the fissioning system will be possible at the FAIR facility in the FELISe campaign. To be able to separate the masses, high resolution detection is required. In particular the time-of-flight should be measured with 35 ps FWHM, which presents a real technical challenge. SOFIA and FELISe experiments will be presented in this work.

Keywords: electromagnetic induced fission, reverse kinematics experiment, nuclear charge determination, nuclear mass determination, GSI, SOFIA, FELISe

1. Introduction

The discovery of nuclear fission in 1938[1,2] has had and continues to have a considerable impact on the evolution of society. This nuclear process, which consists in the splitting of a heavy nucleus into two lighter excited fragments, is characterized by a large energy release (about 200 MeV per fission in the case of fission induced by thermal neutrons on ^{235}U) mainly as kinetic energy. This peculiar property combined with the possibility to initiate chain reactions through the neutrons emitted by the fragments de-exciting in flight, has been very early understood as a formidable potential for energy applications. This initiated a tremendous amount of applied and fundamental research works, which resulted, in particular, into the current nuclear power technology. Also, independently of all applications, the complexity of the fission mechanism and its links to fundamental aspects of nuclear matter has stimulated scientists all over the world to reach a satisfactory understanding capable of accounting for all the facets of the phenomenon.

Huge progress has been accomplished in this regard thanks to experimental investigations and the subsequent theoretical analysis. Nevertheless, no theoretical model is capable to predict quantitatively the fission observables. This is particularly true for the fragments isotopic yields *i.e.* the probability distribution for both the mass A and the atomic number Z of the fragments and the kinetic energies of the individual fragments. Moreover, experimental data are incomplete or inaccurate. In particular, no complete fission fragment isotopic distribution is available, even for the major actinides. Also, cross correlations of the fragments masses, charges and kinetic energies are relatively poorly known.

Various experimental techniques have been used to study fission fragment yields. In most of these cases, experiments were undertaken in direct kinematics and for which nuclear charges have never been completly determined for the whole fission fragment region. Thus, common to all direct kinematics experiments are the limitations in the availability of isotopically separated targets which restrict the experiments to long-lived actinides. This problem was no longer present in the case of the inverse kinematics experiment conducted at GSI Darmstadt.[3] This pioneering work allowed the measurements of charge yields for a large collection of actinides and preactinides. Nevertheless the fission fragment masses were not measured.

The goal of the two fission programs presented in this paper is to complete and give more accurate data on fission fragment properties. Fragments masses, charges and kinetic energies will be measured in coincidence as well as prompt fission neutron multiplicity for the SOFIA short term program, and in addition to this the fission excitation energy will be determined in the FELISe long perspective experiment. Both experiments will take place at the Gesellschaft fur Schwerionen Forschung facility (GSI).

2. Goals of the GSI fission experiments

As mentioned above, experimental data available today on fission yields are incomplete and suffer from a lack of correlations between the different fission observables. This is due to experimental difficulty inherent to the fission reaction. Since the fragments are emitted at a relatively low recoil energy and in opposite directions in the center of mass frame, it is, indeed, uneasy:
(i) to measure with high resolution their kinetic energy, mass or charge,
(ii) to obtain high acceptance and high efficiency measurement,
(iii) to measure simultaneously several parameters e.g. mass, charge and kinetic energy of the fragments.

The GSI fission program is divided into two parts which both use the same principle: electromagnetic induced fission of a relativistic actinide beam. The first one in a short term program called SOFIA (Studies On FIssion with Aladin) which will take place in the current GSI facility in the coming years. Then, Fission@ELISe (FELISe) is a long term program which will benefit from the ELISe instrument from the GSI part to come. Because both are based on inverse kinematics through the use of accelerated beams of fissionable nuclei, these experimental difficulties will be greatly reduced.

SOFIA experiment is expected to start within two years and will enable to determine:
- charge and mass yields with total resolution for both fission fragments in coincidence over the whole domain (*i.e.* no restriction to light or heavy fragments),
- prompt neutron multiplicity for a given fission fragment couple,
- wide range of fissioning nuclei investigated, actinides and preactinides up to mass 238 and nuclear charge 93 for life-time as short as $10\mu s$.

FELISe experiment will enable to determine:
- charge and mass yields with total resolution for both fission fragments in coincidence over the whole domain (i.e. no restriction to light or heavy fragments),
- prompt neutron multiplicity for both fission fragments in coincidence over the whole domain (*i.e.* no restriction to light or heavy fragments),

- wide range of fissioning nuclei investigated, actinides and preactinides up to mass 238 and nuclear charge 93 for life-time as small as some minutes,
- value of excitation energy for each fission.

The conjunction of these data makes this experimental program a real breakthrough in the investigation of nuclear fission. The data obtained will constitute an unprecedented base for applications and fundamental physics and will have a strong impact on nuclear theory and models.

3. The SOFIA experiment

3.1. *The GSI facility*

A ^{238}U beam, extracted from the ion source, is successively accelerated by the UNILAC linear accelerator and the SIS synchrotron up to one GeV per nucleon (fig. 1). This beam impinges a beryllium target in order to

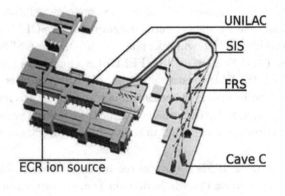

Fig. 1. *View of the GSI facility.*

produce, by fragmentation, radioactive and stable fragmentation products up to the mass 238 and nuclear charge 93. The FRS fragment separator selects through electromagnetic section the selected secondary beam, which are in our case different actinide beams: $^{235-237}$Np, $^{235-238}$U, $^{233-237}$Pa, $^{230-232,234}$Th. Then the SOFIA set-up will be installed in Cave C to benefit from the ALADIN magnet and the LAND neutron wall, which have been successfully used during the past few years. The other types of detectors are still to be built.

3.2. The SOFIA set-up

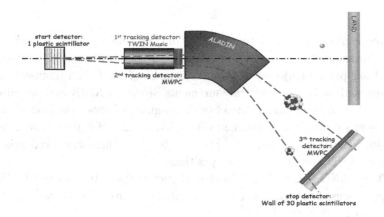

Fig. 2. Schematic view of the SOFIA set-up which will be installed in Cave C.

The actinide fission in an active lead target (fig. 2). The relativistic energy of the actinide beam (600 MeV per nucleon) will lead to a Lorentz contraction of the electromagnetic field of the lead target in the orthogonal direction of the trajectory. The impulsion felt by the actinide is comparable to a gamma absorption (called virtual photon). The energy transferred to the incident actinide will be distributed to all the nucleons. They will enter in resonance before being de-excited by fission. Due to the kinematics of the reaction, both fission fragments are focused in the beam direction within a 40mRad angular cone.

The charge identification is done via a double ionisation chamber, the Twin MUSIC, located after the Pb target. The energy loss of both fragment is detected in each part of the chamber. This measurement, combined with the velocity obtained from the time-of-flight measured between a plastic scintillator inside the active target and a scintillator wall as last detector, gives the ionic charge state of the fragments, therefore their nuclear charge state Z, since all are fully stripped.

Active target, composed by a stack of lead layers, gives information about the layer in which the reaction takes place.

The fragment nuclear mass will be determined by the classical Bρ time-of-flight method. The A/Z ratio of an ion is easily determined using a constant magnetic field (B) by means of the following well known relation:

$$\frac{A}{Z} \sim \frac{B\rho}{\beta\gamma} \tag{1}$$

A precise tracking through the ALADIN magnet combined with the time-of-flight measurement is required in order to get the velocity of the fragments and their precise trajectories. The trajectories of the fission fragments are determined by two position measurements before ALADIN and one after. The first position is extrapolated from the segmented anode twin music. The fragment's position is determined with electron time-of-flight measurement between the cathode and each anode. Then two multiwire proportional chambers will give the two other positions.

The LAND neutron wall, located 10 meters downstream the ALADIN magnet, is used to measure the total number of neutrons evaporated by the fragments.

3.3. Constraints in term of resolution

To be able to separate the masses, high resolution detection is mandatory. The time-of-flight should be measured with 35 ps FWHM, which represents the most difficult part of the project, and in addition to this, the horizontal position resolution needed is 0.2 mm FWHM while the vertical position resolution is 2 mm FWHM.

4. The FELISe experiment

4.1. FAIR

FELISe experiment will be part of the new GSI facility FAIR, Facility for Antiproton and Ion Research. Following the same principle than for SOFIA (fig. 3), a ^{238}U beam is extracted from the ion source and accelerated by the new SIS100 synchrotron up to 1 GeV per nucleon.

By fragmentation in a beryllium target numerous heavy ions are produced, which will be selected by the new separator super FRS before being injected in the New Experimental Storage Ring. FELISe is located on this ring.

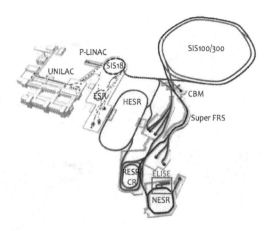

Fig. 3. View of the FAIR facility project.

In the same time, electrons are produced and accelerated up to 500 MeV by the Electron Linear Accelerator and the Eletron Antiproton Ring accelerators. They are then injected in the NESR ring. The fission of actinide is induced by electron by the exchange of a photon between them. Then the electron is scattered and its energy is measured in the electron recoil spectrometer to determine the excitation energy of the fissioning system.

4.2. The FELISe set-up

FELISe set-up follows the same principle than SOFIA set-up (fig. 4). As for the previous experiment, charge identification is done via a double ionisation chamber while mass is given by the classical $B\rho$ time-of-flight method. These fission fragments are deflected by first big dipole. Fragments'position will be measured by three tracking detectors and time-of-flight will be measured between a plastic scintillator and a scintillator wall. Neutron multiplicity will be measured by a neutron wall with a better energy resolution than for LAND detectors to identify from which fragment neutron is coming.

5. Conclusion

SOFIA and FELISe are both very ambitious experimental fission programs. Nuclear charge, mass and kinetic energy will be completly determined for

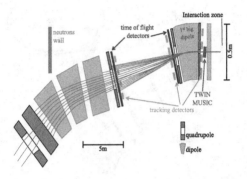

Fig. 4. *Schematic view of the FELSISe set-up.*

each fragment with SOFIA experiment. In addition to this, FELISe will enable to measure neutron multiplicity for each fragment as well as excitation energy of the fissioning system. These data will be measured with a great accury for the fission of a large amount of actinides.

References
1. O. Hahn et al., *Die Naturwissenschaften* **27**, 11 (1939).
2. L. Meitner et al., *Nat.* **143**, 239 (1939).
3. K. Schmidt et al., *Nucl. Phys.* **A 665**, 221 (2000).

THE FISSION FRAGMENT TIME-OF-FLIGHT SPECTROMETER VERDI

S. OBERSTEDT‡, R. BORCEA, F.-J. HAMBSCH, Sh. ZEYNALOV *

EC-JRC Institute for Reference Materials and Measurements, B-2440 Geel, Belgium
‡ *E-mail: stephan.oberstedt@ec.europa.eu*

A. OBERSTEDT, A. GÖÖK †

School of Science and Technology, Örebro University, S-70182 Örebro, Sweden

T. BELGYA, Z. KIS, L. SZENTMIKLOSI, K. TAKÁCS

Institute of Isotopes, Hungarian Academy of Sciences, H-1525 Budapest, Hungary

T. MARTINEZ-PEREZ

Nuclear Innovation Group, Department of Energy, CIEMAT, E-28040 Madrid, Spain

For the investigation of correlated fission characteristics like fragment mass- and energy-distributions the double (v, E) spectrometer VERDI is being constructed. With this instrument we aim at the simultaneous measurement of pre- and post-neutron masses, avoiding prompt neutron corrections. From the simultaneous measurement of pre- and post-neutron fission-fragment data the prompt neutron multiplicity may be inferred as a function of fragment mass and total kinetic energy. In order to arrive at a mass resolving power $\Delta A <$ 2, ultra-fast and radiation hard time pick-up detectors based on artificial diamond material are under investigation. We report about the timing resolution of these detectors obtained with fission fragments, and discuss the first experiment with a single (v, E) version of the VERDI spectrometer performed at the Budapest Research Reactor.

Keywords: Fission fragment distribution; diamond detectors; time-of-flight technique; ^{235}U(n_{th}, f).

---

*Present address: Joint Institute for Nuclear Research, Dubna, Moscow Region, Russia
†Present address: Institute for Nuclear Physics, University of Technology Darmstadt, D-64289 Darmstadt, Germany

1. Introduction

In nuclear fission pairs of fission fragments with different mass and kinetic energy are produced.[1] This process is accompanied by the emission of prompt neutrons and γ-rays from the highly excited fission fragments (FF). The knowledge of the FF yield-distribution, $Y(A, E_k)$, as a function of mass (A) and kinetic energy (E_k) is a key input to fission models and important data for a better understanding the fission process as such. This information may be obtained by means of different techniques. By using a recoil mass separator[2] or measuring the time-of-flight together with the kinetic energy,[3] single fission-fragment yields after prompt neutron emission may be obtained. Employing the double-energy (2E) method, fission-fragment characteristics prior to neutron emission, the so-called pre-neutron masses A* and total kinetic energies, may be obtained, provided prompt neutron emission data are available for the iterative determination of the fragment mass.[4,5] It is just those latter data, which are available only for a few major actinides with sufficient accuracy.

One way to avoid ambiguities introduced in the data analysis by the applied neutron corrections is to measure not only both kinetic energies, but simultaneously their velocities as well. Theoretically, from a double-velocity (2v) measurement both fragment masses before prompt-neutron evaporation may be obtained. The subsequent measurement of the fragment kinetic energies provides the information about the fission-fragment mass after prompt-neutron evaporation. Thus, the difference of pre- and post-neutron masses is just the number of evaporated prompt neutrons. In such a way, those important neutron data do not need to be introduced anymore in the data analysis. Instead, they may be directly deduced as a function of fragment mass and total kinetic energy, from which the excitation energy may be inferred. Such a double time-of-flight (TOF) and energy spectrometer had already been realized in the 1980s at the Institute Laue Langevin and was called Cosi-Fan-Tutte.[3] With an impressive mass resolution ΔA well below 1 mass unit, it suffered from a very small geometrical efficiency of about 4×10^{-5}.[6] Therefore, correlated FF-yield measurements were strongly affected by the non-colinearity of FF emission due to prompt neutron emission. That spectrometer does no longer exist.

Presently, a double (v, E) spectrometer is being built at the JRC-IRMM called VERDI,[7] which aims at a FF mass resolution $\Delta A < 2$ in conjunction with a geometrical efficiency of at least 0.5 %, i. e. about 100 times more efficient than Cosi-Fan-Tutte. In order to achieve a high efficiency, the time pick-up devices should be placed as close as possible to the target under investigation. Since this means that such devices would have to be placed directly in the neutron beam, they have to be radiation resistant. Therefore, an artificial polycrystalline chemical vapor deposited (pCVD) diamond film is considered as detector material to serve as fission event trigger.

In the present work we report about the characterization of polycrystalline CVD diamond detectors (pCVDDD) under irradiation with fission fragments and a first in-situ test of a single (v E) version of the VERDI spectrometer.

2. The Diamond-based Fission Trigger

In nuclear physics diamond detectors are used mainly in high-energy experiments as beam monitors and tracking devices, replacing traditionally employed silicon detectors, because they survive in high radiation environments, have low leakage current and do not need cooling.[8-11] In particular the timing properties of artificial diamonds are remarkable and an intrinsic timing resolution better than 30 ps has been achieved for a mono-energetic ^{52}Cr-beam at incident energy of 650 MeV/u.[12] In view of the properties of this surprising material, it was tempting to see whether a similar timing resolution may be obtained with low energy fission fragments at energies typically between 0.5 and 2.0 MeV/u.

First, the diamond detectors were irradiated with β-particles (the so-called *priming*) before exposed to highly ionizing particles like fission fragments.[13,14] Then, we monitored signal stability up to more than 10^9 fission fragments from a ^{252}Cf source together with the corresponding number of α-particles and fission neutrons for which no signal degradation was observed. Next, we measured the fission-fragment time-of-flight with a symmetric setup consisting of two identical pCVDDDs. The intrinsic timing resolution was then inferred by means of a Monte-Carlo simulation based on published post-neutron data from the reaction ^{252}Cf(SF). The experiment has been performed two-fold: first, using standard analogue electronics and second, using a digital oscilloscope with a bandwidth of 1GHz. The obtained spectra are dipicted in Fig. 1. From various simulations an intrinsic timing resolution $\sigma_{int} < 300$ ps for the analogue and $\sigma_{int} < 150$ ps for the digital measurement was obtained. The timing resolution is comparable with the better micro-channel plate detectors, but diamond detectors are much easier to handle and to operate.

In Fig. 2 the mass distribution from a first single (v, E) measurement is shown. As energy detector a 25 mm^2 silicon detector was used and placed

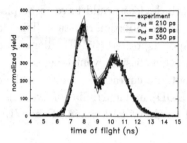 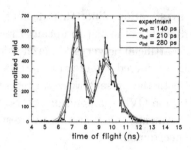

Fig. 1. Fission fragment time-of-flight spectrum for the reaction ^{252}Cf(SF) taken with standard analogue electronics (left) and with a waveform digitizer with a bandwidth of 1GHz (right).

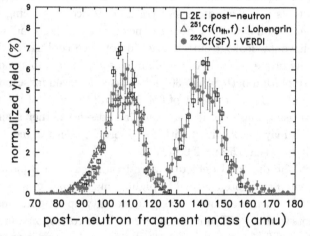

Fig. 2. Experimental fission-fragment mass distribution from the reaction ^{252}Cf(SF) obtained from a TOF measurement with a pCVDDD as fission trigger and a 25 mm^2 silicon-type energy detector, compared to results from Refs.[15–17]

at a distance of 33 cm from the fission source. The data compare well with corresponding ones from the reaction ^{251}Cf(n_{th}, f) measured at LOHENGRIN[15] and obtained from a double-energy measurement.[16,17] The mass resolution, i. e. timing resolution of the spectrometer, still needs to be improved in the future.

3. The Budapest Experiment

At the 10 MW research reactor in Budapest a first in-situ test of a single (v, E) section of the VERDI spectrometer was performed and the post-neutron fission fragments were measured. It contained a ^{235}U sample of mass 113 µg, mounted on a 34 µg/cm^2 thick polyimide backing, a polycrystalline chemical vapour deposited diamond detector(pCVDDD), which provided the fast fission trigger, as well as ten passivated implanted planar silicon (PIPS) detectors for the registration of the fragments' energy and time-of-flight at a distance of 50 cm from the pCVDDD.

The reactor delivered a cold neutron beam with a flux of some 10^7 cm^{-2} s^{-1} resulting in a fission rate above 10^4 s^{-1}. A 5 µg ^{6}LiF sample deposited on a 30 µg/cm^2 thick polyimide backing was mounted directly on the fission sample. The well-known α-triton reaction products were used for online time calibration.

The start signal is induced by an α-particle, a triton or a fission fragment in the pCVDDD, while the signals from the energy detectors provide the stop signal of the coincidence. The fission trigger signal was split and used also for the measurement of prompt fission γ-rays.[18] Time-of-flight and pulse height of the fission fragment were stored in listmode. The experiment was performed during February/March 2010 with effectively 8 days of beam time.

In Fig. 4 a raw fission-fragment time-of-flight spectrum is shown in logarithmic scale. The two Monte-Carlo simulations were performed with two different timing resolutions, 1.0 and 0.4 ns, viz. After correcting the spectrum from the tails, which are due to accidental coincidences with α-particles or tritons, a timing resolution of 400 ps was achieved. Eight out of the ten detectors showed the same behaviour. For 2 detectors the trition energy was below the lower-level of the ADC, preventing an adequate timing calibration. Data analysis and the construction of post-neutron mass-distributions are ongoing.

4. Summary and Conclusions

In this paper we have presented fission time pick-up detectors made from artificial diamond. The timing resolution with standard analogue electronics is well below 300 ps. Using state-of-the-art low-noise broadband electronics this figure may be improved to well below 150 ps as suggested when using a 1GHz waveform digitizer. Radiation hardness was verified for typical fission fragment, α-particle and neutron doses. A first in-situ test of the fission time-of-flight spectrometer VERDI in its single (v, E) configuration demonstrated the suitability of the spectrometer components. At the moment a timing resolution of VERDI as good as 400 ps was deducted from Monte-Carlo simulations, which corresponds to a mass resolution $\Delta A \approx 3$.

Fig. 3. Photograph of the experimental setup with the fission fragment spectrometer VERDI and four lanthanum halide scintillation detectors (upper left part). The neutron beam enters from the right.

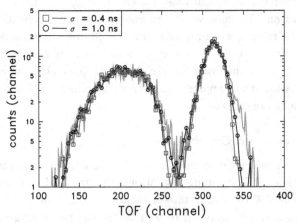

Fig. 4. Raw TOF spectrum obtained from the reaction ^{235}U(n_c, f) together with Monte-Carlo simulations made for two different timing resolutions (see text).

In the next step we will install either a 'transparent' diamond detector or an MCP detector as fission start triger to simultaneously measure both fission fragments.

Acknowledgments

This work was supported by EFNUDAT (agreement number 31027) and NAP VENEUS05 (agreement number OMFB 00184/2006).

References

1. The Nuclear fission process, editor C. Wagemans, CRC Press, ISBN 0-8493-5434-X (1991).
2. E. Moll, H. Schrader, G. Siegert, H. Hammers, M. Asghar, J.P. Bocquet, P. Armbruster, H. Ewald and H. Wollnik, Kerntechnik 8 (1977) 374.
3. A. Oed, P. Geltenbort, R. Brissot, F. Gönnenwein, P. Perrin, E. Aker, D. Engelhardt, Nucl. Inst. Meth. in Phys. Res. 219 (1984) 569.
4. C. Budtz-Jørgensen, H.-H. Knitter, Ch. Straede, F. -J. Hambsch, R. Vogt, Nucl. Inst. Meth. A258 (1987) 209.
5. F. Vivès, F.-J. Hambsch, H. Bax, S. Oberstedt, Nucl. Phys. A662 (2000) 63.
6. N. Boucheneb, M. Ashgar, G. Barreau, T. P. Doan, B. Leroux, A. Sicre, P. Geltenbort and A. Oed, Nucl. Phys. A535 (1991) 77.
7. http://www.fysikersamfundet.se/kf/arsmoten/2009/talks/slides_oberstedt_stephan.pdf
8. RD 42 Collaboration, P. Weilhammer et al., Nucl. Inst. Meth. A 409 (1998) 264.
9. The RD42 collaboration, D. Meier et al., Nuc. Inst. Meth. A 426 (1999) 173.
10. E. Berdermann, K. Blasche, P. Moritz, H. Stelzer, "Diamond Detectors for Heavy-Ion Measurements", XXXVI Int. Winter Meeting on Nuclear Physics, Bormio 1998.
11. E. Berdermann, K. Blasche, P. Moritz, H. Stelzer, B. Voss, F. Zeytouni, Nucl. Phys. B (Proc. Suppl.) 78 (1999) 533-539.
12. E. Berdermann, K. Blasche, P. Moritz, H. Stelzer, B. Voss, Diamond and Related Materials 10 (2001) 1770-1777.
13. M. Marinelli, E. Milani, A. Paoletti, A. Tucciarone, G. Verona Rinati, M. Angelone, M. Pillon, Nucl. Inst. Meth. A 476 (2002) 701-705.
14. S. Oberstedt, C. C. Negoita, T. Atzitzoglu, Scientific Report of the Neutron Physics Unit 2006, EUR Report 23039 EN, Eds. S. Oberstedt, P. Rullhusen, ISBN 978-92-79-05365-8 (2007) 95.
15. E. Birgersson, S. Oberstedt, A. Oberstedt, F.-J. Hambsch, D. Rochman, I. Tsekhanovich, S. Raman, Nucl. Phys. A 791 (2007) 1-23.
16. F.-J. Hambsch and S. Oberstedt, Nucl. Phys. A617 (1997) 347-355.
17. F.-J. Hambsch (2004), private communication.
18. A. Oberstedt, T. Belgya, R. Billnert, R. Borcea, D. Cano-Ott, A. Göök, F.-J. Hambsch, J. Karlsson, Z. Kis, X. Ledoux, J.-G. Marmouget, T. Martinez-Perez, S. Oberstedt, L. Szentmiklosi, K. Takács, proceedings to this conference.

215

DIGITAL PULSE PROCESSING FOR STEFF

A.J. POLLITT*, J.A. DARE, A.G. SMITH, I. TSEKHANOVICH

*School of Physics and Astronomy, The University of Manchester,
Manchester, M13 9PL, The United Kingdom
E-mail: andrew.pollitt@postgrad.manchester.ac.uk

The SpecTrometer for Exotic Fission Fragments (STEFF) is a Manchester based experiment that uses two segmented Bragg Chambers to determine fission fragment energies of a ^{252}Cf source. In this work the pulses collected in the Bragg chambers are processed digitally using Flash ADC's in GRT4 cards and a DAQ. Algorithms have been developed to sort the data pulses off-line. These include routines for the removal of noise and cross talk through digital filtering, adjusting the data for ballistic deficit and adding back pulses from the individual anode segments. From the corrected data, the mass and Z of the fission fragments is deducible. Current methods and present challenges will be discussed.

Keywords: 2v2E Spectrometer; Digital Pulse Processing; Fission Fragment Identification; Pulse Trace Analysis.

1. Introduction

The 1980's saw a revival of ionisation chambers to detect heavy ions. This was due to several factors including their ability to measure the specific ionisation of particles along with their total energy. Therefore it is possible to identify a heavy ion through its mass, A and charge, Z.

One such device that attempted to do this was Cosi Fan Tutte which was a $2v2E$ spectrometer. However the size of the entrance windows to the ionisation chambers was only 15 mm[1] in diameter, which meant that the detector solid angle was small.

In recent years, the study of the nuclear structure of neutron rich nuclei far from stability has been at the forefront of research in the area of nuclear structure physics. Low-energy nuclear fission is able to populate nuclei with a large neutron excess, typically with a ratio $N/Z \sim 1.5$.[2] Therefore performing γ-ray spectroscopy on Fission Fragments is a useful tool.

Combining the ability to measure the A, Z of fission fragments along with the emitted gamma radiation is the objective of a device being built at Manchester. The SpecTrometer for Exotic Fission Fragments (STEFF), is a $2v2E$ device with a larger solid angle than Cosi fan tutte. Using the ability to identify fission fragments by their A and Z, one can reduce the number of detected γ-rays needed to perform γ-ray spectroscopy on fission fragments.

The mass of each of the fission fragments is determined by

$$A = \frac{2E}{v^2}, \qquad (1)$$

where A is the mass, E is the kinetic energy and v is the velocity, of the fragment. The velocity is measured using the time-of-flight of the fission fragment between two timing detectors. The kinetic energy is measured using an ionisation chamber.

This paper will focus on the digital treatment of pulse traces from the STEFF ionisation chambers used to determine both the kinetic energy and the range of the fission fragments. The determination of Z is the subject of the contribution given by I. Tsekhanovich.[3]

2. Spectrometer Design Details

The STEFF setup in the dual-armed, prompt γ-ray configuration can be seen in Figure 1. The fission source used in STEFF for the Manchester based experiment is ^{252}Cf undergoing spontaneous fission. The source is surrounded by 12 NaI(Tl) γ-ray detectors.

Each arm of STEFF consists of 2 timing detectors and an axial ionisation chamber. In the timing detectors the fission fragments pass through a thin piece of aluminised mylar, liberating electrons. These electrons are accelerated through a high voltage and are deflected through 90 degrees where they are detected by microchannel plate and a Multi-Wire Proportional Counter (MWPC) in the start and stop detectors respectively.

The ionisation chamber has an entrance window of dimensions 25 cm × 15 cm, made from 0.5 μm thick Mylar. The chamber is filled to a pressure of ~60 mbar with isobutane gas in order to stop the fission fragments of ^{252}Cf within the space between the front wall of the chamber (cathode), and the Frisch Grid used to shield the anode.

The Frisch grid is situated 15.6 cm from the window, in the intervening space are field shaping rings, used to keep the electric field in this region uniform. The anode is segmented into 15 rectangular segments which are operated at a voltage of ~1200 V. Each anode segment is connected to the

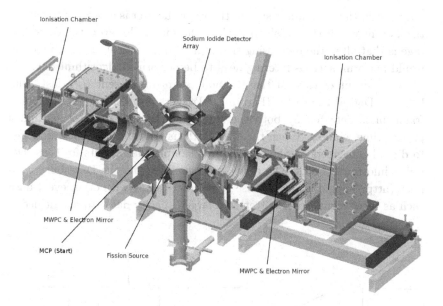

Fig. 1. A cross sectional diagram of the SpecTrometer for Exotic Fission Fragments in the dual armed prompt γ-ray configuration. Showing the positions of the start detector, stop detector and axial ionisation chamber.

high voltage supply, and a charge sensitive pre-amplifier (CSPA) through a decoupling capacitor.

3. Digital Pulse Processing

3.1. *Digital Electronics*

In the Cosi Fan Tutte spectrometer, analogue electronics was used to measure the ionisation chamber pulses.[4] The kinetic energy of the fission fragments was determined by integrating the total charge deposited in the ionisation chamber. The range of the fission fragments in the chamber was determined by the time difference between the fragment entering the chamber and the output pulse reaching a certain fraction of the overall expected pulse amplitude.

STEFF employs digital electronics to record the entire pulse trace received by an ionisation chamber anode segment, these are then analysed off-line. The advantages of using digital electronics in such a way is: that one can set a suitable range threshold after determining the overall pulse amplitude and the raw data is stored, therefore the user is at liberty to change the way in which they wish to analyse the pulse. However, the disadvantage is that all of the processing that standard nuclear electronics modules would perform, such as filtering need to be performed algorithmically.

The output of each of the 15 anode segments is sent to a fast Analogue to Digital Converter (Figure 2). Triggers are created when a fission fragment is detected by both the start and the stop detectors within a pre-set time window. When the VME BUS is triggered the flash analogue to digital converters on the GRT4 cards sample their inputs for 6 μs. Detailed information on GRT4 cards is obtainable from the Daresbury webpage (http://npg.dl.ac.uk). These pulse traces along with other event data such as γ-ray signals are written to a large capacity data storage device.

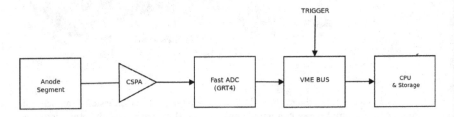

Fig. 2. A block diagram to show the way in which pulse traces from the ionisation chamber are collected by the data acquisition system.

3.2. Data Processing

Data are sorted off-line, Figure 3 shows 100 sequential raw pulse traces of all energies without any algorithmic treatment. The only requirement on those pulses is that they are full energy pulses in the central anode. In a paper by Oed et al,[5] pulse traces from the Cosi Fan Tutte ionisation chamber were presented. For heavy gasses (like Xenon), $1/40^6$ of these pulses shown bumps and distortions which inhibited the measurement of the range of the fission fragments. Over 7000 raw pulse traces, in different energy bins from STEFF have been inspected and the bumps and distortions of the kind presented were not observed.

The ionisation chamber pulses have been inverted and offset by the GRT4 cards and are suffering from ballistic deficit due to the finite pre-amplifier decay time. A correction is applied to the pulses for the ballistic deficit by calculating the deficit using the known pre-amplifier decay time

and adding it back to the pulse incrementally. The correction also removes
the baseline and inverts the pulse trace.

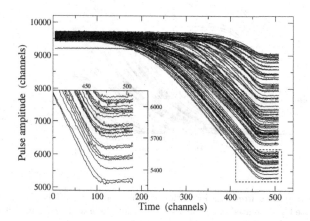

Fig. 3. 100 sequential raw pulse traces from the ionisation chamber. The only condition
on these pulses is that they are full energy hits in the central anode. These pulses are
suffering from ballistic deficit, they have also been inverted by the GRT4 card and have
a baseline offset. Ballistic deficit gives the pulse trace above channel ~ 475 a gradient
where it should be flat. Note the absence of any distorted traces.

Noise filtering is also done algorithmically, firstly, a high pass filter is
applied to remove any low frequency noise. This differentiates the pulses
over a range of about of 10 channels. It then re-integrates them which has
the effect of smoothing the pulses on the scale of 10 channels. A low pass
6 pole Chebyshev filter[7] is also applied to the pulses to remove any high
frequency noise. After noise filtering the signal to noise ratio for light fission
fragments is $\sim 0.08\%$, comparable to the level usually given by analogue
electronics modules. Examples of these traces are shown in Figure 4.

Electronic drift is detected with a pulser signal. A drift match coefficient is calculated by comparing the channel number of the pulser and the
position that the pulser should be at. This is applied to the real signals. In
practice this correction is very small.

The Add-Back algorithm is used to sum the contributions of real pulses
from different anode segments together. When the algorithm detects a
pulse, above a certain threshold, it looks geometrically at the anode seg-

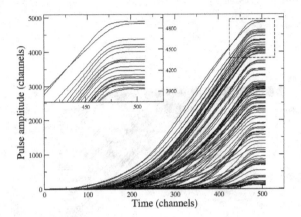

Fig. 4. 100 pulse traces from the ionisation chamber after they have been inverted, baseline subtracted and treated with the algorithms for Ballistic Deficit and Noise. The signal to noise ratio for the largest pulses is ~ 0.08%.

ments surrounding it. If a surrounding anode has a pulse with a height greater than the noise level, it is considered to be part of the real signal. All of the contributions are summed together to create the overall pulse trace.

The energy loss algorithm performs a correction to compensate for the energy loss, ΔE, of fission fragments as they pass through the ionisation chamber window. Simulations were performed using SRIM[8] for the fractional energy loss for the Fission Fragments of ^{252}Cf with the highest yield passing through the window. This was performed on fragments with masses, $A = 75$ to 166, covering the majority of the mass distribution. The algorithm uses the results of these simulations to iteratively work back to the original energy. This is still under development.

4. Conclusion

The SpecTrometer for Exotic Fission Fragments is a $2v2E$ spectrometer in the latter stages of development at Manchester. STEFF employs digital electronics to determine the kinetic energy and the range of fission fragments in its axial ionisation chambers. Several algorithms for the digital treatment of pulse traces from the axial ionisation chambers have been discussed. These include algorithms for the removal of noise through digital filtering, adjusting the data for ballistic deficit and adding back pulses from the individual anode segments.

5. Acknowledgements

This work was supported by the Engineering and Physical Sciences Research Council in The United Kingdom, grant number 2-4570.5.

References

1. A. Oed, P. Geltenbort, F. Goennenwein, T. Manning and D. Souque, *Nuclear Instruments and Methods in Physics Research* **205**, 1455 (1983).
2. I. Ahmad and W. R. Phillips, *Reports on Progress in Physics* **58**, p. 1415 (1995).
3. I. Tsekhanovich, Progress in Atomic Number Identification of Fission Products, in 7th *Seminar on Fission*, (Gent, Belgium, 2010).
4. A. Oed, P. Geltenbort and F. Goennenwein, *Nuclear Instruments and Methods in Physics Research* **205**, 451 (1983).
5. A. Oed, P. Geltenbort and F. Goennenwein, *Radiation Effects* **96**, 53 (1986).
6. F. Goennenwein, Personal Communications (2010).
7. S. W. Smith, *The Scientist & Engineer's Guide to Digital Signal Processing*, 1st edn. (California Technical Pub, 1997).
8. J. Ziegler, SRIM 2008 (2008), Computer Programme.

FISSION γ-RAY MEASUREMENTS WITH LANTHANUM HALIDE SCINTILLATION DETECTORS

A. OBERSTEDT*, R. BILLNERT, J. KARLSSON, A. GÖÖK

School of Science and Technology, Örebro University, S-70182 Örebro, Sweden
*E-mail: andreas.oberstedt@oru.se

S. OBERSTEDT, F.-J. HAMBSCH

EC-JRC Institute for Reference Materials and Measurements, B-2440 Geel, Belgium

X. LEDOUX, J.-G. MARMOUGET

CEA/DAM Ile de France, F-91297 Arpajon, France

T. BELGYA, Z. KIS, L. SZENTMIKLOSI, K. TAKÁCS

Institute of Isotopes, Hungarian Academy of Sciences, H-1525 Budapest, Hungary

T. MARTINEZ-PEREZ, D. CANO-OTT

Nuclear Innovation Group, Department of Energy, CIEMAT, E-28040 Madrid, Spain

In a recent experiment, performed at the 10 MW research reactor at the Institute of Isotopes (IKI) in Budapest, the emission of prompt γ-rays from the cold-neutron induced fission of ^{236}U* was measured. For that purpose four cerium-doped lanthanum halide scintillation detectors were employed and found very useful in order to distinguish between γ-rays from different reactions. Although data analysis is not completed yet, we could show that these novel detectors indeed provide the means towards new and more precise input data necessary for the modeling of γ-heating in nuclear reactors.

Keywords: Prompt fission gamma-rays; Lanthanum halides; Scintillation detector; Time-of-flight technique; ^{235}U(n_{th}, f).

1. Introduction

Requests for new measurements on prompt γ-ray emission in the reactions ^{235}U(n_{th}, f) and ^{239}Pu(n_{th}, f) have been formulated and included in the Nuclear Data High Priority Request List of the Nuclear Energy Agency (NEA, Req. ID: H.3, H.4).[1] However, a major difficulty in such measurements is, apart from the need to obtain a sufficient mass resolution for fission fragments, the clear suppression of background γ-rays induced by prompt fission neutrons in the γ-detector. A usual method here is to discriminate γ-rays and neutrons by their different time-of-flight. The quality of discrimination is strongly coupled to the timing resolution of the detector, which is usually not better than 5 ns for NaI detectors. A promising step towards better data might be achieved by using the recently developed cerium-doped lanthanum halide scintillation detectors, which have shown to provide a timing resolution better than 500 ps[2,3] together with a more than 40 % better energy resolution compared to NaI, i.e. less than 4 % (FWHM) compared to 6.5 % at 662 keV (^{137}Cs).[4]

2. The Lanthanum Halide Detectors

As mentioned above four lanthanum halide detectors were used to measure prompt γ-rays from the reaction ^{235}U(n_{th}, f). In more detail, two coaxial LaCl$_3$:Ce detectors of size 1.5 in. × 1.5 in.[5] as well as one with dimensions 3 in. × 3 in.[6] were used. The fourth detector consisted of a cylindrical LaBr$_3$:Ce crystal[6] with diameter and length of 2 in. each.

Prior to the experiment presented in this paper, the first mentioned detectors have been characterized, which is described elsewhere.[7] They were found to have an intrinsic timing resolution of about 440 ps, measured with a ^{60}Co source. The energy resolution was determined for γ-energies from 81 to 6919 keV, exhibiting more or less the expected $E^{-1/2}$ behavior. For 662 keV (^{137}Cs) the obtained energy resolution was around 4 % (FWHM), in agreement with values provided by the manufacturer.[5] Together with a dynamical range up to more than 11 MeV, these detectors showed a good linearity with residuals far below 1 %. The full intrinsic peak efficiency was observed to be 53 % better than the one for NaI:Tl detectors of the same size.[8]

The properties of the larger LaCl$_3$:Ce detector were not fully known at the time of the experiment, its efficiency, however, could be expected to be more than three times higher compared to the smaller detectors around 1 MeV,[9] mostly for geometrical reasons. The LaBr$_3$:Ce detector was delivered just before the start of the experiment and was, thus, not characterized yet. Nevertheless, from previous studies it is known that these detectors are even better than LaCl$_3$:Ce detectors in terms of both energy and timing resolution, with about 2.8 % (FWHM) at 662 keV and coincidence resolving times of < 300 ps.[3]

Fig. 1. Photograph of the experimental setup with four lanthanum halide scintillation detectors (upper left part) and the fission fragment spectrometer VERDI. The neutron beam enters from the lower edge.

3. The Budapest Experiment

At the 10 MW research reactor in Budapest, γ-rays were measured in coincidence with fission fragments, which were detected by using the fission fragment spectrometer VERDI. It contained a ^{235}U sample of mass 113 μg, mounted on a 34 μg/cm^2 thick polyimide backing, a polycrystalline chemical vapour deposited (pCVD) diamond detector, which provided the fast fission trigger, as well as ten passivated implanted planar silicon (PIPS) detectors for the registration of the fragments' energy and time-of-flight. The details about that part of the experiment were topic of a dedicated presentation at this conference.[10] The reactor delivered a cold neutron beam with a flux of some 10^7 cm^{-2} s^{-1}, causing a fission rate above 10^4 s^{-1}. The four scintillation detectors were placed outside the time-of-flight spectrometer VERDI at a distance of about 30 cm (cf. Fig. 1). Blankets containing ^{6}Li and lead blocks were applied as shielding against scattered thermal neutrons and γ-rays, respectively. The signals from the four γ-detectors were fed into constant fraction discriminators (CFD) and further into the same time-to-analog converter (TAC) of range 1 μs. They were giving the start signal, while the signals from the diamond detector, with an appropriate delay, provided the stop signal of the coincidence. For all four scintillation detectors the pulse height was stored in listmode, together with three pulse

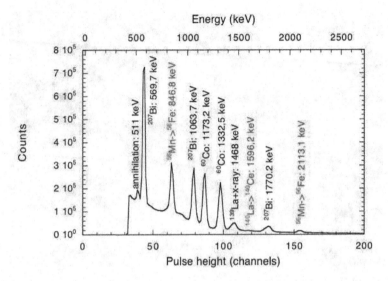

Fig. 2. Pulse height calibration spectrum of the large $LaCl_3$:Ce detector. The different colours of the energy assignments refer to γ-peaks from the known sources ^{60}Co and a ^{207}Bi, and from activation in the detector and its environment, respectively.

shape discrimination signals for the $LaCl_3$:Ce detectors only, as well as the TAC signal. The experiment was performed during February/March 2010 with ten days of actual beam time and below we present and discuss the first preliminary results.

4. Results and Discussion

Prior to and right after the beamtime the detectors were energy calibrated using different radioactive sources. Figure 2 shows the calibration spectrum of the large $LaCl_3$:Ce detector, taken with a ^{60}Co and a ^{207}Bi source after neutron irradiation of the target. The γ-peaks are assigned to the corresponding source and energy, the assignments in a lighter colour, however, belong to peaks that were not observed before the neutron beam was opened. They were identified to correspond to activation and successive decay of atoms in the crystal by the reaction $^{139}La(n_{th},\gamma)$ and induced by fission neutrons through the reaction $^{56}Fe(n,p)^{56}Mn$ in the wall of the vacuum chamber of VERDI, respectively. Apart from that, we noticed that the conversion from pulse height to energy practically did not change over the period of 10 days of experiment. Moreover, from the counts under the 1596.2 keV peak, the known volume and efficiency of the detector we could

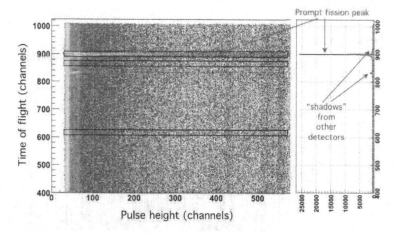

Fig. 3. Two-dimensional representation of γ-rays detected with the large LaCl$_3$:Ce detector by their TAC-signal versus pulse height. To the right a projection on the TAC-axis is shown, indicating the prompt fission γ-rays together with some corresponding events from other detectors. The three rectangles denote areas, where prompt fission γ-rays (upper) and γ-rays from fast neutron (middle) and thermal neutron (lower) induced reactions are observed.

estimate the thermal neutron flux impinging on the detectors to be less than 0.2 cm^{-2} s^{-1}.

For each detector the data treatment was carried out as follows:

- first a selection was made on the pulse shape signal (where available) in order to discriminate the γ-rays from other incident particles, including α-particles from the intrinsic activity of the crystals
- then the dependence of the TAC-signal from pulse height was determined and corrected for
- different windows were set in order to distinguish events from different γ-ray producing reactions, i.e. prompt fission as well as thermal and fast neutron induced process such as (n_{th}, γ), (n, n') and (n, p)

An example for the results of this procedure is shown in Fig. 3. There, the rectangular regions are indicating in which TAC-signal regime the different processes were expected. During data analysis, however, these windows were chosen as large as possible. For each of them pulse height spectra were generated and energy calibrated and the processes were verified by identifying the characteristic γ-spectra. The thermal-neutron induced γ-rays (including the ones from the activation of VERDI) were not correlated with fission fragments and thus independent from time. They were consid-

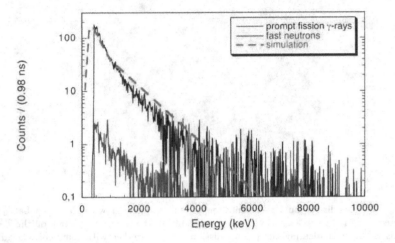

Fig. 4. Normalized, energy-calibrated and background-subtracted spectra for the 3 in. × 3 in. LaCl$_3$:Ce detector from both inelastic scattering of fast neutrons and prompt decay of fission products together with the result of simulations.

ered as background, normalized per TAC-channel and subtracted from the other spectra. Also the γ-rays from inelastic neutron scattering were subtracted from the prompt γ-rays, which results in the energy distribution of prompt fission γ-rays. Figure 4 shows normalized, energy-calibrated and background-subtracted spectra for the large LaCl$_3$:Ce detector from both inelastic scattering of fast neutrons and prompt decay of fission products. To guide the eye, the result of a simulation[11] is included and compared to the experimental fission γ-spectrum, which is described by

$$N(E) = \begin{cases} 38.13 \cdot (E - 0.085)e^{1.648E} & E < 0.3 MeV \\ 26.8 \cdot e^{-2.30E} & 0.3 < E < 1.0 MeV \\ 8.0 \cdot e^{-1.10E} & 1.0 < E < 8.0 MeV \end{cases} \quad (1)$$

as obtained from previous measurements.[12] Although the experimental spectrum is not corrected for efficiency yet and only a part of the taken data is analyzed so far, the good agreement is promising. What remains to be done is the analysis of all data as well as determining the response functions for all detectors and applying them to the spectra.

5. Summary and Conclusions

In this paper we have presented first and preliminary results from the recent measurement of prompt γ-rays from the reaction $^{235}U(n_{th}, f)$. Despite the early stage of data analysis the comparison with a simulation of the energy distribution appears encouraging. This makes us optimistic to provide eventually new and more precise data to resolve present deficiencies in γ-ray production data in evaluated nuclear data files. The employment of fast detectors like lanthanum halide detectors in conjunction with pCVD diamond detectors will account for that.

Acknowledgments

This work was supported by EFNUDAT (agreement number 31027) and NAP VENEUS05 (agreement number OMFB 00184/2006).

References

1. http://www.nea.fr/html/dbdata/hprl/hprlview.pl?ID=421 and http://www.nea.fr/html/dbdata/hprl/hprlview.pl?ID=422.
2. K. S. Shah, J. Glodo, M. Klugerman, L. Cirignano, W. W. Moses, S. E. Derenzo, M. J. Weber, Nucl. Instr. Meth. A 505 (2003) 76.
3. A. Iltis, M. R. Mayhugh, P. Menge, C. M. Rozsa, O. Selles, V. Solovyev, Nucl. Instr. Meth. A 563 (2006) 359.
4. B. D. Milbrath, B. J. Choate, J. E. Fast, W. K. Hensley, R. T. Kouzes, J. E. Schweppe, Nucl. Instr. Meth. A 572 (2007) 774.
5. SCIONIX Holland bv, P.O. Box 143, 3980 CC Bunnik, The Netherlands.
6. Saint-Gobain Crystals, 17900 Great Lakes Pkwy, Hiram, OH 44234-9681, USA.
7. A. Oberstedt, R. Billnert, J. Karlsson, S. Oberstedt, W. Geerts, and F.-J. Hambsch, in: A. Chatillon, H. Faust, G. Fioni, and D. Goutte, H. Goutte (Eds.), Fourth Int. Workshop on Nucl. Fission and Fission-Product Spectroscopy, AIP Conf. Proc., vol. 1175, 2009, p. 257.
8. G. Gilmore, Practical Gamma-ray Spectroscopy, John Wiley & Sons, ISBN 978-0-470-86196-7 (2008).
9. R. Nicolini, F. Camera, N. Blasi, S. Brambilla, R. Bassini, C. Boiano, A. Bracco, F. C. L. Crespi, O. Wieland, G. Benzoni, S. Leoni, B. Million, D. Montanati, A. Zalite, Nucl. Instr. Meth. A 582 (2007) 554.
10. S. Oberstedt, T. Belgya, R. Borcea, A. Göök, F.-J. Hambsch, Z. Kis, T. Martinez-Perez, A. Oberstedt, L. Szentmiklosi, K. Takács, Sh. Zeynalov, proceedings to this conference.
11. J. M. Verbeke, C. Hagmann, D. Wright, UCRL-AR-228518, Lawrence Livermore National Laboratory (2009).
12. F. C. Maienschein, R. W. Peelle, T. A. Love, Neutron Phys. Ann. Prog. Rep. for Sept. 1, 1958, ORNL-2609, Oak Ridge National Laboratory (1958).

CHARACTERIZATION AND DEVELOPMENT OF AN ACTIVE SCINTILLATING TARGET FOR FISSION STUDIES

G. BELIER, J. AUPIAIS

CEA, DAM, DIF
F-91297 Arpajon, France

Actinide targets are of prime importance for our laboratory, in the context of neutron induced fission. We are currently developing a new kind of active target for such studies. The goal is to use commercially available scintillating organic liquid used for α spectrometry. These liquids contain extractive molecules that allow to dissolve any actinide isotopes in the liquid itself providing very high detection efficiency. The development includes the fabrication of a dedicated system for photon counting that can be used as an active target in nuclear experiments. Fission triggering and total kinetic energy measurements are targeted. The scintillator response to fission has been characterized, thanks to the ^{252}Cf spontaneous fission. It comprises energy response, pulse shape discrimination capability and quenching with respect to the actinide concentration. Simulations were performed in order to characterize the detection efficiency and neutron spectra distortion for neutron induced fission experiments.

1. Introduction

Fast neutron induced reactions are important for future nuclear plant generations, since most of the retained solutions are fast reactors. In this context fast neutron induced fission is of prime importance and one is faced to the lack of data together with incomplete understanding of the process. Studies with fast neutrons are especially difficult because the available fluxes are much lower than for facilities delivering thermal neutrons and at the same time the cross sections are lower than at thermal energies. The consequence is that higher masses are needed to perform experiments. This is already true for cross sections measurements but it becomes crucial in the case of prompt neutron or gamma spectra measurement or for experiments were several observables are correlated.

From a technical point of view, working with high masses has several disadvantages. For some isotopes with high specific activities it leads to high α activity with an associated important pileup effect. High masses also means large sample surfaces and large target sizes and masses. One of the consequences is that the target capacitance for a fission chamber containing 100 mg of actinide is high and lead to an electronic signal to noise ratio degradation. Associated to a consequent pileup effect this lead to a bad α-fission discrimination and poor time resolutions.

In this presentation we report the development of a new kind of active target. The principle is to dissolve an actinide into an organic liquid scintillator coupled to a photo-detector. The advantages of such a target are the ease of fabrication, the high attainable actinide concentrations, the pulse shape discrimination capability of organic scintillator, the fast time response and finally the possibility to reprocess the actinide in case of detector failure.

After a brief presentation of the liquid scintillation technique, the results obtained with the spontaneous fission of ^{252}Cf will be detailed. It includes the response to fission fragment (both the energy and pulse shape were characterized), the α-fission discrimination and the light quenching against actinide concentration. Simulations of neutron spectra distortion will be presented and considerations on the detection efficiency will be given. Finally we will conclude with perspectives on future developments on this active target, and on experiments that will be performed with.

2. PERALS technique

The Photon Electron Rejection Alpha Liquid Spectrometry (PERALS) is commonly used for α spectrometry. The review and developments realized were described in a D.O.E. report by W.J. McDowell [1]. In this work we use the actinide extracting scintillator ALPHAEX from ORDELA, Inc. Company. The solvent is toluene (7.5 mol/L). The fluorescent molecule is PBBO (10^{-2} mol/L) used with naphthalene (1.5 mol/L) as both light shifter and α/β discrimination enhancer and the extracting molecule is HDEHP (0.2 mol/L). The respective hydrogen and carbon concentrations are 0.52 and 0.47 at.%. Such scintillator presents a rather good energy resolution which is around 200 keV for natural uranium α lines and allows the use of pulse shape discrimination (PSD) techniques in order to reject recoiling electrons.

3. Characterization of the response to fission fragments

To our best knowledge the response of this scintillator to fission fragment is almost unknown. Only one energy spectrum obtained with ^{252}Cf was published in the context of Super Heavy Element search at GSI [2], but information on fission response is scarce. In the present work ^{252}Cf spontaneous fission was also used. The experiments were performed by using Pyrex tubes inserted into a mechanical device containing a light reflector and a quantacon Burle 8850 photomultiplier associated to an ORTEC 265 base. The photomultiplier voltage was -1800 V. The optical link was obtained with a silicone oil (Ordela Part no. LC-01). For the pulse shaping and shape discrimination a Mesytec MPD4 module was used. This module implements the zero crossing time technique for Pulse Shape discrimination (PSD) and delivers a Time to Amplitude Converter (TAC) signal. The shaped signal amplitude and the TAC were coded with a 7072 FAST Inc. ADC, associated to a FAST Inc. data acquisition system. The pulse height amplitude and TAC parameters were used to build a bidimensional histograms (PSD%PHA histogram).

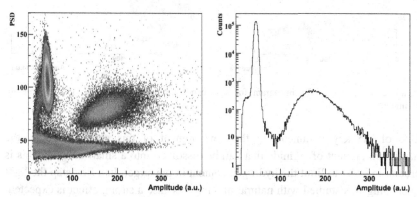

Figure 1. Left: Bidimensional histogram with the Pulse height Amplitude on the x axis and the Pulse Shape Discrimination (PSD) TAC parameter on the y axis. Right: corresponding β suppressed pulse height amplitude spectrum.

The left part of Figure 1 shows an example of histogram performed with a 1 mL liquid scintillator solution traced with 7 Bq of ^{252}Cf. Three classes of event are clearly seen. The long tailed one is associated to electrons. The located one at small amplitude and large TAC values is associated to the ^{252}Cf α decay. Finally the last one at intermediate TAC values is associated to the ^{252}Cf spontaneous fission. It clearly demonstrates the capability of this detector to discriminate fission from over events.

The right part of Figure 1 presents the beta suppressed pulse height amplitude (PHA) spectrum. It can be calibrated by using the known 6.118 MeV ^{252}Cf α line and the mean averaged fission total kinetic energy which is 188 MeV. The resulting width at half maximum for the fission bump is 78 MeV for an expected value of around 30 MeV. This result clearly indicates that regarding the energy response further experiments are needed.

4. Light quenching against actinide mass

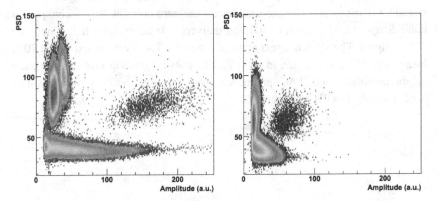

Figure 2. Left: PSD%PHA histogram for a U concentration of 10 µg/cm^3. Right: same for 10 mg/cm^3.

One of the key parameter in the development of this active target is the maximum amount of actinide that can be dissolved into a small volume. This is crucial in order to limit the hydrogen content in the target. In a first step the light quenching was studied with natural uranium because a strong effect is expected. Three solutions were prepared with concentrations of 10 µg/cm^3, 1 and 10 mg/cm^3. Figure 2 shows the comparison of the histograms obtained for the 2 extreme concentrations. A strong light attenuation is observed on the pulse height amplitude parameter while the pulse shape TAC is less affected. Nevertheless for the highest uranium concentrations, a simple fission trigger with a good α-fission discrimination can still be obtained.

The same study was performed with natural thorium which from a quenching point of view is representative of every actinide except uranium. Figure 3 presents the same histograms as in Figure 2 and for the same concentrations. No quenching effect is measured nor on the amplitude nor on the pulse shape. Hence for most actinide one can envisage to obtain higher concentrations of up to 100 mg/cm^3. Additional quenching from specific chemicals and radiolysis aging for high specific activities will have to be studied. Signal pile-up will be also a limiting factor on the maximum attainable concentration.

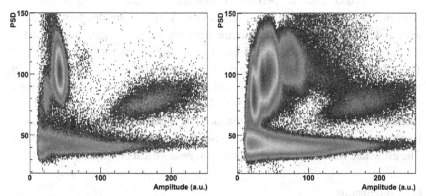

Figure 3. Same as Figure 2 but for thorium.

5. Simulations

Two kinds of simulations were performed. The first one was done in order to study the neutron spectrum distortion due to scattering on hydrogen contained in the target. The simulations consist in counting the number of fission of ^{235}U against the energy of the neutron that produced the reaction. This isotope is disadvantageous because fission can occur at any energy. The proportions of fission events that occur at energies lower than the incident neutron energy are respectively 6.7%, 3% and 1.7% for respectively 1, 5 and 10 MeV initial energies. The geometry used for this simulation is the usual PERALS geometry (Figure 4). With a dedicated system, flatter geometry can be used in order to minimize distortions. Tests were done on a dedicated system whose geometry can be changed as needed. It is composed of an aluminium container which is painted to ensure light reflection, and a pure quartz window. The photo-detector is an avalanche photodiode which can provide better energy resolutions than a photomultiplier [3] together with a very compact target. The first tests show that the use of a fast preamplifier developed for pulse shape discrimination with silicon detectors [4] can be used for the same purpose with the scintillator.

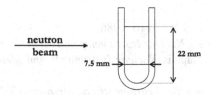

Figure 4. Geometry adopted for neutron spectrum distortion simulations.

The second kind of performed simulations were devoted to detection efficiency calculations. The count losses in such system are due to wall effects; with a skin whose thickness equals the detected particle range. It is around 50 µm for 5 MeV alpha particles and between 20 and 30 µm for fission fragments. The simulation for alpha particles gave a detection efficiency loss of 0.34%. This value is in agreement with the published measurement [1]. In the case of fission, the simulations are not finalized yet. A much lower loss is expected due to smaller ranges, but mainly due to the fact that 2 particles are emitted in fission. Hence the only losses correspond to fission axis tangent to the liquid surface. The estimated maximum loss obtained from geometrical considerations is 10^{-6}.

6. Conclusions

Very encouraging results were obtained from the characterization of this new kind of active target. Fission triggering and pulse shape discrimination were demonstrated. Further work will be performed on the energy response of the scintillator in order to do Total Kinetic Energy measurements in fission. Experiments on spontaneous and neutron induced fission will be performed with the use of a high efficiency 4π neutron long counter. In these experiments the total gamma ray energy can also be measured with this tank detector. Pulse shape discrimination can also be improved by using the fast/slow charges method that can also be used for light ion identification. This last aspect together with the very high efficiency might allow to measure new rare heavy ion disintegrations or to improve the data for such decays. Contrary to experiments using a mass spectrometer the active target enable to discriminate heavy ion decay and spontaneous fission thanks to the big difference in the total kinetic energies of both processes. Hence this exotic decay can be distinguished from ternary fission. Nevertheless the cluster emission is most intense in the region of doubly magic nucleus ^{208}Pb. Extremely low rates are expected for actinides and fission count losses have to be minimized. Interesting results might be obtained with dedicated large volume cells linked to photomultipliers.

References

1. W.J. McDowell, NAS-NS-3116, 1986.
2. B. Wierczinski et al., *Nuclear Instruments and Methods* **A370** (1996) 532.
3. A. Reboli and J. Aupiais *Nuclear Instruments and Methods* **A550** (2005) 593.
4. H. Hamrita et al., *Nuclear Instruments and Methods* **A531** (2004) 607.

Fission Probabilities: Barriers and Cross Sections

THERMAL FISSION CROSS SECTION MEASUREMENTS OF ^{243}CM AND ^{245}CM

L. POPESCU, J. HEYSE, J. WAGEMANS,
SCK•CEN, Boeretang 200, B-2400 Mol, Belgium

C. WAGEMANS
University of Ghent, Proeftuinstraat 86, B-9000 Ghent, Belgium

A new measurement program was set up at SCK•CEN to determine the thermal neutron-induced fission cross section of a number of Cm isotopes. The experiments are performed at a thermal neutron beam from the graphite moderated reactor BR1 at SCK•CEN. This paper presents preliminary results of our ^{243}Cm(n,f) and ^{245}Cm(n,f) cross-section measurements.

1. Introduction

Neutron-induced fission cross-section data are of interest for basic nuclear physics and astrophysics studies and are a crucial input in reactor physics calculations. Spent nuclear fuel contains important quantities of Cm isotopes. Hence Cm(n,f) cross-section data are needed for nuclear waste transmutation calculations. They play an important role in the core design of Accelerator Driven Systems and Generation IV reactors.

The currently available evaluated databases show important gaps and discrepancies. Some of these evaluations are based on old measurements using poorly characterized samples with low enrichment. Therefore, new measurements are required to improve the evaluated nuclear data libraries.

In the framework of the Interuniversity Attraction Poles project financed by the Belgian Science Policy Office of the Belgian State, a measurement program was set up at SCK•CEN to determine the thermal neutron-induced fission cross section of a number of Cm isotopes.

2. Experimental setup

The measurements were performed at the Z59 irradiation channel of the graphite moderated BR1 reactor [1]. The neutron beam has been collimated to a diameter of 20 mm by a system of three boron carbide collimators. The samples and the detectors have been installed in a vacuum chamber, with the samples positioned perpendicular to the beam, at 60 cm from the reactor wall. At the sample position the thermal neutron flux was of the order of $6 \cdot 10^5$ $s^{-1}cm^{-2}$.

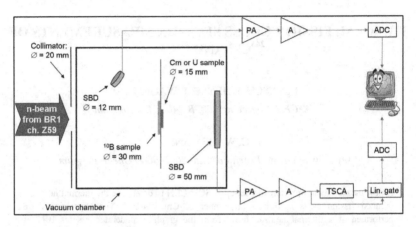

Figure 1. Schematic view of the detection system. The samples and the two surface barrier detectors (SBD) have been mounted in a vacuum chamber. The data acquisition system is sketched on the right-hand side of the picture.

The Cm(n,f) measurements have been performed relative to ^{235}U(n,f). Two samples have been mounted in the center of a vacuum chamber. They were positioned back-to-back on a support plate perpendicular to the beam (see Fig. 1). One of these samples was either ^{235}U, or the fissile material under investigation; the other was a ^{10}B sample, used for monitoring the neutron flux. The ^{10}B nuclei in the monitoring sample are subject to (n,α) reactions with a high cross section; therefore, the sample ensures an accurate determination of the neutron flux variations.

The detection of the emitted alphas and fission fragments was done via surface barrier detectors (SBD) positioned as shown in Fig. 1. The collected signals were pre-amplified, amplified, passed through an ADC module and then collected by a mixing unit before being sent to the acquisition PC. A linear gate was inserted after the amplifier in the Cm chain, to cut most of the pile-up signals caused by α decay of nuclei in the Cm sample.

3. ^{243}Cm(n,f) measurement and results

A Cm-oxide sample containing (2.320±0.025) µg of ^{243}Cm was used for the ^{243}Cm(n,f) measurement. The material was deposited on a 30 µm thick Al backing. The sample diameter was 15 mm and its isotopic composition at the moment of the experiment (April 2010) is given in Table 1.

Table 1. Isotopic composition of the ^{243}Cm sample.

Isotope	Number of Atoms	At.%
^{243}Cm	5.70x10^{15}	92,503
^{244}Cm	2.57x10^{14}	4,173
^{245}Cm	4.01x10^{12}	0,065
^{246}Cm	5.72x10^{13}	0,928
^{247}Cm	1.34x10^{12}	0,022
^{248}Cm	1.70x10^{12}	0,028
^{239}Pu	1.41x10^{14}	2,282

The measured ^{243}Cm(n,f) pulse height spectrum is plotted in Fig. 2. As the sample is a strong α-emitter, the pile-up of several α-signals can produce a signal with amplitude close to the one of the fission fragments. This undesirable background (represented by the large peak at the left side of the spectrum) could be well separated from the ^{243}Cm(n,f) region.

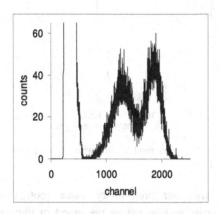

Figure 2. Pulse height spectrum measured with the ^{243}Cm sample. The signals produced by the fission fragments could be well separated from the α pile-up (the large peak on the left side of the figure).

As the even-A isotopes in the sample decay also via spontaneous fission (SF), we determined the SF contributions to the ^{243}Cm(n,f) spectrum by closing the neutron beam and repeating the measurement. The result indicates almost 16% contributions from SF to the spectrum shown in Fig. 2. This spectrum contains also a 9.5% contribution from the epithermal component of the neutron beam, which was determined by repeating the ^{243}Cm(n,f) measurements with a

Cd filter (1 mm thick) positioned perpendicular to the beam, in front of the vacuum chamber.

For determining the neutron flux and the detection efficiency, a ^{235}U(n,f) measurement was performed keeping the same configuration as for ^{243}Cm. A 99.94% enriched ^{235}U sample, with a diameter of 15 mm and a mass of (193.29±1.33) µg was used. This sample was deposited on a 30 µm thick Al backing. For the ^{235}U(n,f) cross section we used the $\sigma_f(E_n = 25.3 \text{ meV}) = 582.60$ b value as given in Ref. [2].

The ratio of the counting rates for the ^{10}B(n,α) measurement performed in parallel with the ^{235}U(n,f) and ^{243}Cm(n,f) measurements allowed an accurate normalization of the beam flux. In the present analysis, the Westcott g_f-factors, at $E_n = 25.3$ meV, $g_f(^{235}\text{U}) = (0.9771 \pm 0.0008)$ and $g_f(^{243}\text{Cm}) = 1.0054$ as given in Ref. [2] have been used to correct for the deviation from the 1/v shape of these fission cross sections.

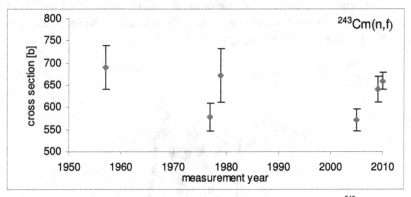

Figure 3. Comparison of the present result (2010) with literature data for the ^{243}Cm(n,f) cross section at thermal energy [3-7].

Figure 3 compares our preliminary value (660±18) b with different ^{243}Cm(n,f) thermal cross section values presented in literature [3-7]. Some of these authors did not consider the Westcott factor when extrapolating the cross section to $E_n = 25.3$ meV [3,6]. Moreover, we did not renormalize the old values using the latest cross sections for the calibration measurement (e.g. ^{235}U(n,f)).

The thermal fission cross section adopted in the European (JEFF-3.1.1), Japanese (JENDL-AC-2008) and American (ENDF/B-VII.0) neutron libraries are 617.4 b, 587.36 b and 613.32 b, respectively. Our result overlaps within the error bars with the results of Hulet [3], Zhuravlev [5] and Alekseev [7] and it shows a deviation of 7% from the value adopted by JEFF-3.1.1 and 12% from the value adopted by JENDL nuclear data library.

The statistical uncertainty of the present result is less than 1%, while the systematic uncertainty is almost 2%. The main systematic uncertainty is induced by the uncertainty on the mass of the sample.

4. ^{245}Cm(n,f) measurement and results

For the ^{245}Cm(n,f) measurement, we used a Cm-oxide sample containing (141±3) μg of ^{245}Cm, which was deposited on a 30 μm thick Al backing. The sample diameter was 15 mm and its isotopic composition at the date of the experiment (November 2008) is given in Table 2.

Table 2. Isotopic composition of the ^{245}Cm sample.

Isotope	Number of Atoms	At.%
^{244}Cm	3.16x10^{15}	0.898
^{245}Cm	3.48x10^{17}	98.687
^{246}Cm	1.43x10^{15}	0.406
^{247}Cm	2.82x10^{13}	0.008
^{248}Cm	7.06x10^{12}	0.002

The measured ^{245}Cm(n,f) pulse height spectrum is plotted in Fig. 4. Similarly to the case of ^{243}Cm, the sample is a strong α-emitter, the pile-up of several α-signals producing a signal with amplitude close to the one of the fission fragments. This background could be well separated from the ^{245}Cm(n,f) region.

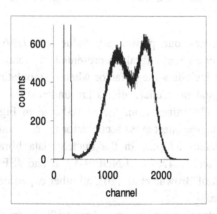

Figure 4. Pulse height spectrum measured with the ^{245}Cm sample. The signals produced by the fission fragments could be well separated from the α pile-up (the large peak on the left side of the figure).

The SF contributions and the contributions from the epithermal component of the neutron beam to the ^{245}Cm(n,f) spectrum shown in Fig. 4 were determined in the same way as for the case of ^{243}Cm(n,f) measurement. The SF contributions were of the order of 1.6%, while the contributions from the epithermal component were of the order of 2.6%.

A value of g_f (^{245}Cm) = (0.954 ± 0.033) [2] has been used in the data analysis to correct for the deviation from the 1/v shape of the fission cross section. The large uncertainty in the Westcott g_f-factor for ^{245}Cm induces a comparatively large systematic uncertainty in the obtained ^{245}Cm(n,f) thermal cross section.

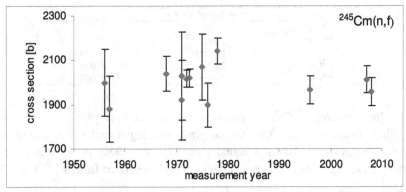

Figure 5. Comparison of the present result (2008) with literature data for the ^{245}Cm(n,f) cross section at thermal energy [8-18].

Figure 5 compares our preliminary value (1927±86) b with different ^{245}Cm(n,f) thermal cross section values presented in literature [8-18]. Most of these values are not including the g_f-factor when extrapolating the cross section to E_n = 25.3 meV and their uncertainties are underestimated. Similarly to the case of ^{243}Cm, the old values from the plot shown in Fig. 5 have not been renormalized by using the latest cross sections for the calibration measurements.

The thermal values adopted in the nuclear data libraries are 2054.1 b (JENDL-AC-2008) and 2142.4 b (ENDF/B-VII.0 and JEFF-3.1.1). However, apart from the result of Browne et al. [16], all other experimental data point to a smaller cross section.

Our result overlaps within the error bars with the previous results and it shows a deviation of about 10% from the value adopted by the ENDF/B-VII.0 and JEFF-3.1.1 nuclear data libraries and 6.2% from the value adopted by JENDL-AC-2008.

The statistical uncertainty of the present result is of the order of 0.4% while the systematic uncertainty is 4.1%. As already mentioned, the main systematic uncertainty is induced by the large uncertainty on the g_f-factor for ^{245}Cm. Recent measurements at the GELINA facility of IRMM, Geel, lead to a more accurate value: g_f (^{245}Cm) = (0.939±0.019) [19]. This new result leads to a value of (1957±65) b for the thermal cross section of ^{245}Cm(n,f), which has a better overlap with previous results. It shows a deviation of 8.6% from the value adopted by the ENDF/B-VII.0 and JEFF-3.1.1 nuclear data libraries and 4.7% from the value adopted by JENDL-AC-2008.

5. Conclusions

The ^{243}Cm(n,f) and ^{245}Cm(n,f) thermal cross sections at 25.3 meV have been measured at the BR1 reactor of SCK•CEN. Our preliminary analysis indicate a thermal cross-section value of (660±18) b for ^{243}Cm and (1927±86) b for ^{245}Cm. Although these preliminary results deviate up to 12% from the values adopted by different nuclear data libraries, they overlap within error-bars with most of the previous experimental values [3-18]. A better agreement is obtained for ^{245}Cm, when using the recent g_f value of Serot et al.[19], which leads to a thermal cross-section value of (1957±65) b.

Acknowledgments

This work was performed in the framework of the Interuniversity Attraction Poles project financed by the Belgian Science Policy Office of the Belgian State.

References

1. J. Wagemans, "The BR1 reactor: a versatile tool for fission experiments" in proceedings of the "*Seminar on Fission*" (2007).
2. S. F. Mughabghab, "Atlas of Neutron Resonances: Resonance Parameters and Thermal Cross Sections Z=1-100", Publisher: Elsevier Science (2006).
3. E. K. Hulet, R. W. Hoff, H.R. Bowman and M. C. Michell, *Phys. Rev.* **107**, 1294 (1957).
4. C. E. Bemis Jr., J. H. Olivier, R. Eby and J. Halperin, *Nucl. Sci. Eng.* **63**, 413 (1977).
5. K. D. Zhuravlev and N.I. Kroshkin, *Atom. En.* **47**, 55 (1979).
6. O. Serot, C. Wagemans, S. Vermote, J. Heyse, T. Soldner and P. Geltenbort, *AIP Conf. Proc.* **798**, 182 (2005).
7. A. A. Alekseev et al., *Atom. En.* **107**, 86-90 (2009).
8. P. R. Fields et al., *Phys. Rev.* **102**, 180 (1956).
9. E. K. Hulet et al., *Phys. Rev.* **107**, 1294 (1957).

10. H. Diamond et al., *J. Inorg. Nucl. Chem.* **30**, 2553 (1968).
11. M. C. Thompson et al., *J. Inorg. Nucl. Chem.* **33**, 1553 (1971).
12. J. Halperin et al., *"Chemical Division Annual Progress Report 1970"*, ORNL 4581, p. 28 (1971).
13. R. W. Benjamin, K.W. MacMurdo and J.D. Spencer, *Nucl. Sci. Eng.* **47**, 203 (1972).
14. K. D. Zhuravlev, N. I. Kroshkin and A. P. Chetverikov, *Atom. En.* **39**, 285 (1975).
15. V. D. Gavrilov et al., *Atom. En.* **41**, 185 (1976).
16. J. C. Browne, R. W. Benjamin and D. G. Karraker, *Nucl. Sci. Eng.* **65**, 166 (1978).
17. M. I. Kuvshinov et al., Conference on the Physics of Reactors, Mito, Japan 1996, vol. 2, page 338, (1996).
18. O. Bringer et al., "Measurements of thermal fission and capture cross sections of minor actinides within the Mini-INCA project" in proceedings of the *"International Conference on Nuclear Data for Science and Technology"* (2007).
19. O. Serot et al., *these proceedings*.

^{245}CM FISSION CROSS SECTION MEASUREMENT IN THE THERMAL ENERGY REGION

O. SEROT[*]

CEA-Cadarache, DEN/DER/SPRC/LEPh, Bat. 230, F-13108 Saint Paul lez Durance, France

C. WAGEMANS, S. VERMOTE

Department of Physics and Astronomy, University of Gent, B-9000 Gent, Belgium

J. VAN GILS

EC-JRC Institute for Reference Materials and Measurements, B-2440 Geel, Belgium

A new cross section measurement for the ^{245}Cm(n,f) reaction in the thermal energy region has been performed at the GELINA neutron facility of the Institute for Reference Materials and Measurements (IRMM) in Geel, Belgium. The energy of the neutrons is determined applying the time of flight method using a flight path length of about 9 m. In the present work, the incident neutron energy covers 10 meV up to a few eV. A 98.48% enriched ^{245}Cm sample was mounted back-to-back with a ^{10}B sample in the centre of a vacuum chamber together with two surface barrier detectors positioned outside the neutron beam. One detector measured the ^{10}B(n,α)^{7}Li reaction products for the neutron flux determination, while the second one registered the ^{245}Cm(n,f) fragments. In this way, the neutron flux can be determined simultaneously with the fission fragments. A control measurement has been performed replacing the ^{245}Cm sample with a ^{235}U sample in order to check that the well-known ^{235}U(n,f) cross section can be reproduced. Our measurement yielded a ^{245}Cm(n$_{th}$,f) cross section of 2131±43±173 b and a Westcott factor g_f=0.939±0.019.

1. Introduction

In order to improve the prediction of heavy actinide concentrations in reactor fuel elements and to reduce the long-term nuclear waste radiotoxicity by using transmutation, cross sections for neutron-induced reactions are required by nuclear industry for many minor actinides, in particular for the ^{245}Cm(n,f) reaction. In the resolved resonance region, only a few old measurements exist for this reaction: Browne's measurement [1] which covers the thermal region up to 35 eV, Moore's one [2] which starts at 20 eV up to 3 MeV and White's one [3]

---
[*] Corresponding author. E-mail address: olivier.serot@cea.fr

from 10 eV up to 63 eV. Recently, the ^{245}Cm fission cross section has been measured by Calviani [4] at the n_TOF facility (CERN), but the results are not yet available in the EXFOR database and will not be used in the present paper for comparison. In addition, for the ^{245}Cm thermal neutron-induced fission cross section, a strong dispersion between measurements is observed as reported by Popescu et al. [5]. This dispersion is partly due to the poor knowledge of the so-called Westcott factor, which describes the deviation of the cross section from a 1/v shape. Up to now, this Westcott factor has been determined from Browne's measurement which is the only one performed in the thermal energy region by the time of flight (TOF) procedure. In this context, a new measurement of the ^{245}Cm(n,f) cross section has been performed between 0.01 eV up to 2 eV.

2. Experimental setup

The measurement was carried out at the GELINA neutron time of flight facility of the Institute for Reference Materials and Measurements in Geel (Belgium). This accelerator delivers a pulsed and compressed electron beam that hits a rotating U target, hence producing bremsstrahlung gamma rays. These gamma rays create neutrons via (γ,n) and (γ,f) reactions. The neutrons are then moderated in 4 cm thick water-filled beryllium containers. The length of the flight path is 9.3 m.

The neutron-induced fission cross section was measured using two different repetition frequencies of the accelerator: 800 Hz and 50 Hz. With 800 Hz, a cadmium filter must be placed in front of the chamber in order to remove overlapping neutrons from previous bursts. So, the incident neutron energy range is in between 0.5 eV up to a several hundred keV. Preliminary results obtained with this repetition frequency have been already published elsewhere [6]. With a 50 Hz repetition frequency, overlapping neutrons from two consecutive bursts do not occur anymore, and therefore the cadmium filter can be removed. In this way, the thermal energy region can be investigated.

The curium sample was mounted in the centre of a vacuum chamber, back-to-back with a ^{10}B layer as shown in Fig. 1. Two Si-Au surface barrier detectors, positioned outside the neutron beam, are used. The first one (fission side) detects the fission fragments from the (n,f) reactions while the second one (boron side) registers alpha particles coming from the ^{10}B(n,α) reactions. In this way, the neutron flux can be determined simultaneously with the fission fragments. These detector signals are then amplified, digitized, and stored in a personal computer.

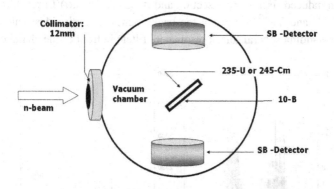

Figure 1. Schematic view of our experimental setup used for the ^{245}Cm neutron-induced fission cross section measurement.

The measurement is performed in two separate runs. For the first one, we use a highly enriched U sample. It permits us to determine the detection geometry factors on both sides and to check if the well known ^{235}U(n,f) cross section can be reproduced. For the second run, the U sample is replaced by the curium one, maintaining the same detection geometry.

3. Measurement with the 235-U sample

The fission fragments counting rate measured on the fission side (Y_f) can be written as follows:

$$Y_f - Y_f^{BGR} = 0.5 \times \varepsilon_f N_U \varphi(E_n) \sigma_f (E_n) \qquad (1)$$

The ^{235}U(n,f) pulse-height spectrum integrated over all incident neutron energies is plotted in Fig. 2 (left part).

With the discriminator setting used on the boron side, only the alpha particles are detected (see Fig. 2, right). Hence, the measured counting rate Y_B is given by:

$$Y_B - Y_B^{BGR} = \varepsilon_B N_B \varphi(E_n) \sigma_B (E_n) \qquad (2)$$

In eqs. (1) and (2), N_U and N_B are the number of uranium and boron atoms, $\varphi\square(E_n\square)$ is the neutron flux, ε_B and ε_f are the detection geometry factors, σ_f is the neutron induced fission cross section and σ_B is the $^{10}B(n,\alpha)^7Li$ reaction cross section. Y_f^{BGR} and Y_B^{BGR} represent the backgrounds on both sides which were measured by adding several 'black resonance' filters in front of the chamber.

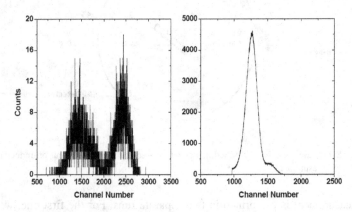

Figure 2. Pulse height spectra measured on the uranium (left) and boron (right) sides.

From the above equations and using the JEFF3.1 cross sections for σ_f and σ_B, it is possible to extract the $\varepsilon_f / \varepsilon_B$ ratio. Then, the $^{235}U(n,f)$ cross section can be deduced by applying the following equation:

$$\sigma_f(E_n) = \frac{\varepsilon_B}{\varepsilon_f} \frac{N_B}{N_U} \frac{0.5(Y_f - Y_f^{BGR})\sigma_B^{JEFF31}(E_n)}{(Y_B - Y_B^{BGR})} \quad (3)$$

Results are shown in Fig. 3 and are compared with the JEFF-3.1 evaluation file. The very nice agreement which could be achieved gives confidence in our energy calibration and in the good functioning of our experimental setup.

4. Measurement with the 245-Cm sample

The thickness of the curium oxide sample is 50 µg/cm². The active diameter is 15 mm. The isotopic composition in March 2003 is given in Table 1.

Table 1. Isotopic composition of the ^{245}Cm sample.

Isotope	244Cm	245Cm	246Cm	247Cm	248Cm
Abundance	1.11	98.48	0.405	0.008	0.002

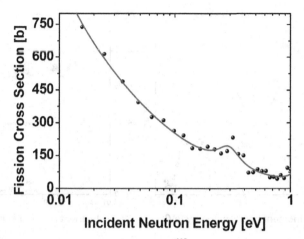

Figure 3. Neutron-induced fission cross section of ^{235}U (circle) compared with the JEFF3.1 evaluation (line).

As described for the ^{235}U measurement, the background measurement was performed by adding several 'black resonance' filters in front of the chamber: Cd, Rh, Au, Mn, W and Co. The low energy part of the background spectra are plotted in Fig. 4. On the fission side (right part of the figure), the observed strong background is due to spontaneous fission events from ^{244}Cm and ^{246}Cm. The corresponding time of flight spectrum is also plotted, showing, as expected, a flat behavior. Then, the ^{245}Cm(n,f) cross section can be determined applying eq. (3). Preliminary results obtained in this way are plotted in Fig. 5 from 0.01 eV up to 2 eV. Results deduced from the 800 Hz run [6] and the JEFF3.1.1 evaluation are also shown in the figure. Our measured thermal neutron-induced fission cross section yields:

$$\sigma_f(E_n = 0.0253\text{eV}) = (2131 \pm 43 \pm 173)\,\text{b}$$

where the first mentioned uncertainty (43 b) corresponds to the statistical uncertainty, whilst the second one (173 b) is an estimation of the systematic errors. Our thermal fission cross section value is in good agreement with Browne's one [1]. From our experimental data, we have also calculated the Westcott g_f factor [7], defined as:

$$g_f(T) = \frac{2}{\sqrt{\pi}\sqrt{E_0}\,E_T^{3/2}\sigma_f(E_0)} \int_0^\infty \sigma_f(E)\,E\,\exp(-\frac{E}{E_T})\,dE$$

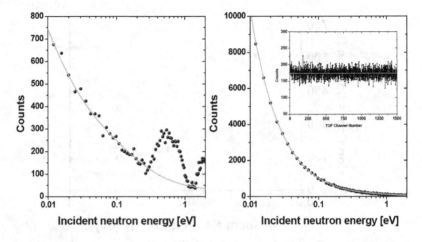

Figure 4. Background measurements obtained with 'black resonance' filters on the boron (left) and fission (right) sides. On the fission side, the background is entirely due to the ^{244}Cm and ^{246}Cm spontaneous fission, yielding a flat behavior of the TOF spectrum (see insert).

Using T=293.6K and $E_0=E_T=25.3$meV, we have found: $g_f=0.939\pm0.019$, which can be compared with the recommended value proposed by Mughabghab [8]: $g_f=0.954\pm0.033$.

5. Conclusion

The neutron-induced fission cross section of ^{245}Cm has been measured at the GELINA facility in the thermal energy region. From a preliminary analysis of our experimental data, a new thermal fission cross section and a new Wescott factor were determined. Combining the present measurement with the one in the resonance energy region will lead in a near future to a new set of resonance parameters.

Acknowledgments

Part of this work was performed in the framework of the Interuniversity Attraction Poles project financed by the Belgian Science Policy Office of the Belgian State.

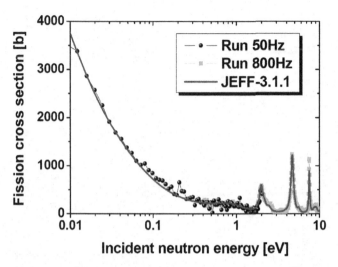

Figure 5. Preliminary results of the neutron-induced fission cross section of ^{245}Cm in the thermal energy region (circles). Data in the resonance energy region (squares) from Ref. [6] are also plotted together with the JEFF3.1.1 evaluation.

References

1. J.C. Browne, R.W. Benjamin, D.G. Karraker, *Nucl. Sci. Eng.* **65** 166 (1978).
2. M.S. Moore and G.A. Keyworth, *Phys. Rev.* **C3** 1656 (1971).
3. R. M. White et al., *Int. Conf. on Nucl. Cross Sections F. Techn.*, Knoxville 1979, p.496.
4. Calviani et al., *Proc. of the 3rd Int. Conf. on Fission and Fission Product Spectroscopy*, Cadarache (France), May 2009, p. 211.
5. L. A. Popescu, J. Heyse, J. Wagemans, C. Wagemans, *these proceedings.*
6. O. Serot, C. Wagemans, S. Vermote and J. Van Gils, *Proc. of Int. Conf. on Nuclear Data for Science and Technology* (ND-2010), Jeju Island (Korea), April 26-30, 2010, in press.
7. C. Westcott, *J. Nucl. Energy*, **2**, 59 (1955).
8. S. F. Mughabghab, *Atlas of Neutron Resonances*, Elsevier Science, Amsterdam, 2006.

r-PROCESS REACTION RATES FOR THE ACTINIDES AND BEYOND

I. V. PANOV* and I. Yu. KORNEEV

Institute for Theoretical and Experimental Physics,
Moscow, 117259, Russia
*E-mail: igor.panov@itep.ru
www.itep.ru

T. RAUSCHER and F.-K. THIELEMANN

Department of Physics, University of Basel,
CH-4056 Basel, Switzerland
E-mail: F-K.Thielemann@unibas.ch

We discuss the importance of different fission rates for the formation of heavy and superheavy nuclei in the astrophysical r-process. Neutron-induced reaction rates, including fission and neutron capture, are calculated in the temperature range $10^8 \leq T(K) \leq 10^{10}$ within the framework of the statistical model for targets with the atomic number $84 \leq Z \leq 118$ (from Po to Uuo) from the neutron to the proton drip-line for different mass and fission barrier predictions based on Thomas-Fermi (TF), Extended Thomas-Fermi plus Strutinsky Integral (ETFSI), Finite-Range Droplet Model (FRDM) and Hartree-Fock-Bogolyubov (HFB) approaches. The contribution of spontaneous fission as well as beta-delayed fission to the recycling r-process is discussed. We also discuss the possibility of rate tests, based on mini r-processed yields in nuclear explosions.

Keywords: Fission; Nuclear reactions; Nucleosynthesis; Abundances; Superheavy elements.

1. Introduction

Investigations of the r-process make use of reaction networks including thousands of nuclei and tens of thousands of reactions. Most of these reactions occur far from stability and thus cannot be directly studied, yet, in the laboratory. This is especially true for the region of fissionable nuclei, which is the focus of the present report.

Fission has often been neglected in astrophysical calculations. In early applications to astrophysical nucleosynthesis, usually only one mode was considered, beta-delayed fission[1] or a phenomenological model of spontaneous fission[2–4]. However, it was shown recently that neutron-induced fission can be more important than beta-delayed fission in r-process nucleosynthesis[5–7].

When considering the main r-process component where the r-process can proceed beyond the transuranium region and possibly reach also the region of superheavy elements (SHE), all decay modes become important. That has been the main motivation for extending and uptodating neutron-induced fission and (n,γ) cross section predictions[8] up to the heaviest chemical elements for which mass and fission barrier predictions exist[9–14]. These data are needed for r-process scenarios in high neutron density environments and to evaluate the question whether superheavy elements (SHE) can be formed in the r-process. Beta-delayed fission[1,15–17] as well as spontaneous fission are important in the late stages of the r-process[6], will also be considered in future investigations.

These new calculations of neutron-induced reaction rates[18], expanded deeply into the transuranium region, can be tested by utilizing the only existing experiments with measured yields of transuranium elements observed in thermonuclear explosions in the 20th century[19–22].

2. Neutron-induced fission rates

As in previous approaches[8,23,24] we have applied the statistical Wolfenstein-Hauser-Feshbach formalism[25,26] for the calculation of neutron-induced cross sections and reaction rates.

In addition to (n,γ)-reactions the fission channel was also included as it was described in[1,16,24,27]. The statistical model is applicable for astrophysical rate calculations as long as there is a sufficiently high density of excited states in the compound nucleus at the relevant bombarding energy, which is the case for most heavy nuclei.

The fission transmission coefficient $T_f(E, J^\pi)$ includes the sum over all possible final states and is evaluated as discussed in[16,24,28] and is related to the fission probability $P_f(E, J^\pi) = T_f(E, J^\pi)/T_{\text{tot}}(E, J^\pi)$ considered in the papers cited above.

When making use of a double-humped fission barrier[29], the fission transmission coefficients can be calculated in the limit of complete damping, which averages over transmission resonances, assuming that levels in the second minimum between the first and second barrier E_A and E_B are equally spaced. The individual barriers can be approximated by (inverted

parabola) Hill-Wheeler barrier shapes.
Ground state spins and parities are known for many unstable nuclei. Far off stability, theoretical ground state spins and parities[30] were taken when experimental values were unavailable. At excitation energies where the level information is not complete anymore a nuclear level density[31] was used. This method is based on the back shifted Fermi-Gas approach, where the level density parameter a and back shift δ are obtained globally from the appropriate mass model employed.

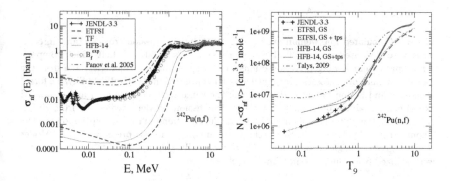

Fig. 1. Comparison of energy-dependent experimentally based (JENDL-3.3)[32] (n,fission)-cross-sections (left panel), and (n,f)-rates, integrated over Maxwell-Boltzmann distributions of targets and projectiles for the displayed temperatures, (right panel) for ^{242}Pu. The symbols are chosen as follows: TALYS-predictions[13] (green two-dot-dash line), JENDL-3.3[32] (crosses). Our calculations are shown for reaction rates including only the ground state (GS) or a thermally populated target (GS+tps) for different nuclear data predictions (HFB or ETFSI) and experimental data, marked B_f^{exp} as well. Only the GS rates can be compared to experimental data. The results of[13] correspond to GS + tps conditions.

In Fig. 1 we compare our predictions for neutron-induced fission cross sections as well as the calculations of neutron-induced fission rates for ^{242}Pu with evaluated neutron data from JENDL-3.3[32,34] and other predictions. (The accuracy of the evaluated data is usually not specified. In the region of interest the accuracy of up-to-date measurements for plutonium isotopes varies from 2% to 15%[35] and is not larger than the plot signs.)

An extended comparison to evaluated neutron-induced fission cross sections for the trans-lead region is shown in Fig. 2. The left panel shows the difference between the calculated ones when the HFB mass and fission barrier predictions were used: in this work (triangles down) and by Talys (crosses). In the right panel of Fig. 2 the calculated cross sections made use

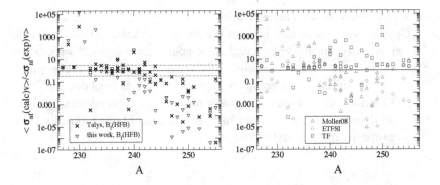

Fig. 2. Calculated maxwellian averaged neutron-induced fission cross sections in comparison to evaluated experimental fission cross sections at 30 keV. Left panel: comparison of maxwellian averaged cross-sections from our calculations (triangles) and from the Talys[13] calculations (crosses), both utilizing HFB predictions (with BSk14 Skyrme force) of masses and fission barriers. Right panel: the same ratio of calculated and evaluated experimental values, but employing theoretical fission barrier predictions from different sources: the Extended Thomas-Fermi model by Mamdouh et al.[10] (diamonds), the Thomas-Fermi by Myers and Swiatecki[12] (squares), and recent predictions from Moller et al.[33] (green triangles up). When not experimentally known, consistent nuclear masses were utilized from the corresponding model predictions.

of mass and fission barrier predictions from TF and ETFSI as well as new predictions[33]. The differences in cross section predictions emerge from the differences in fission barrier heights employed, as indicated in the legend, different level density used etc. The utilization of predictions by Möller et al.[33] probably gives a better agreement for the heaviest isotopes.

3. How to test the rates?

The predictive power of the extrapolations to isotopes with high neutron excess reflects the gross behavior of mass predictions (which is also related to the evaluation of fission barrier heights). While r-process applications and the comparison to solar abundances are one way of testing the nuclear input, there exist also experiments with thermonuclear explosions from the 1960's with observed abundance yields[20–22]. Here we made an attempt to utilize the opportunity to test the calculated rates and from different mass and fission barrier predictions.

In the past, when detailed cross section predictions were not available, yet, a number of simplified evaluations and model predictions were used to obtain neutron rates, based on one given cross section in an isotopic chain. This included linear extrapolations with the ratio of encountered neutron capture Q-values,[36] which were applied successfully[15] in obtaining fits to explosion yields. Another extrapolation model[37] was based on averaged cross-section systematics for $\sigma_{n\gamma}$, reflecting more consistently the dependence of the level density at the excitation energy in the compound nucleus[38].

In the present application we make use of our statistical model predictions of fission rates for different mass and fission barrier predictions[16,18]. All the theoretical predictions[13,18] display a strong dependence on mass and especially fission barriers[18], but they are the step forward in fission rate calculations in comparison with phenomenological models often used before.

We have used the simple one-energy-group model of neutron irradiation of the target, consisting of ^{238}U with admixtures of some other Pu or Np isotopes. The exposure time as well as the neutron density are the parameters, which vary in the range[39]: $1 \times 10^{-7}s < \tau < 3 \times 10^{-7}s$ and $n_n \sim$ Flux/v $\sim 5 \times 10^{23}$ cm^{-3}.

We introduced this toy-model just to evaluate whether the knowledge from actinides yields of observed thermonuclear explosions can test our rate predictions, depending on the initial composition and the time profiles of density, neutron number and energy. We have made a number of runs with different initial composition, based on different suggestions[20,21,39], and confirmed that the observed yields can be explained only if the nucleosynthesis started on a mixed initial seed nuclei composition, or if the irradiation followed a variation in fluxes as a function of time[36] (see also Fig. 3).

Our numerical calculations lead us to the fact that the initial composition should be known at least roughly. We could show that the main behavior of the observed abundance curve can be explained by 1) one target material but being exposed to varying fluxes of different durations - in agreement with earlier evaluations[36]. 2) a superposition of yields from different target materials in a one-energy-group approximation, which probably existed in target[20,21]. 3) a combination of options 1) and 2). For this reason we choose for our presentation of rate tests the target mixture of U, Pu and Np isotopes just to show the both, the strong influence of odd-even effects in capture rates combined with fission processes.

Already from the calculations in such a simple model it is clear, that due to the mass predictions[9-12,14] we have seen a strong odd-even effect, which is the smallest in case of HFB[13] masses. A change in the odd-even effect is clearly seen, when beta-delayed fission is included in addition to neutron-induced reactions (Fig. 3, right panel). We have to point out that the changes of odd-even effects appearing in the yield-curve, can be explained by beta-delayed fission. Probably the theoretical models overestimate the effect for nuclei with A< 250.

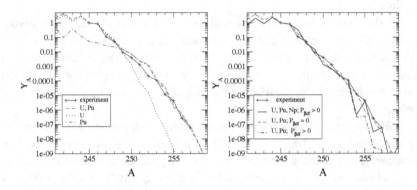

Fig. 3. Left: Y_A dependencies of Pu-mixture to ^{238}U target, when fission was not included. τ=150ns. Right panel: Influence of fission processes on Y_A. Basic input: (n,γ) and (n,fission) are considered as well as Pu+U target. Crosses - experimental data[19-21]. HFB mass and fission barrier predictions were used.

4. Conclusions and discussion

The calculations of neutron-induced reaction rates - neutron capture and neutron-induced fission are made for the extended number of nuclei involved in the r-process up to Z=118, for the temperature range $10^8 \leq T(K) \leq 10^{10}$ and a variety of mass and fission barrier predictions based on Thomas-Fermi (TF), Extended Thomas-Fermi plus Strutinsky Integral (ETFSI), Finite-Range Droplet Model (FRDM) and Hartree-Fock-Bogolyubov (HFB) approaches. Extended tables as well as the temperature dependent fits are published in AA and available at CDS via http://cdsweb.u-strasbg.fr/cgi-bin/qcat?J/A+A.

The modelling of an impulse r-process, which occurs in thermonuclear explosions, provides the opportunity to test the predicted neutron-induced and fission rates in comparison to observed actinide yields. As one conclusion of such a test it was shown that rates calculated on the basis of HFB-predictions lead to a reduction of the odd-even effect for neighboring isotopes. The numerical calculations confirmed that a mixture of some additional isotopes to the uranium target (Pa, Np, Pu) were either formed during early stages of the explosion or existed in the initial sample. We clarified on the basis of numerical calculations that the inversion of the odd-even effect for isotopes with $A > 250$, observed in experiment, should be explained by beta-delayed fission. We also indicate that the calculated beta-delayed probabilities are probably overestimated at least for isotopes with $A < 250$.

For more detailed tests of our reaction rate predictions a better known initial composition of the target and the time-dependent behavior of the density and neutron flux are desiable.

Acknowledgments

The authors thank A. Kelić-Heil, Yu. S. Lutostansky, V. I. Lyashuk, G. Martínez-Pinedo, Yu. Ts. Oganessjan and K.-H. Schmidt for useful discussions. This work was supported by the Swiss National Science Foundation (SCOPES project No. IZ73Z0-128180/1) and grant 2000-105328. I.P. and I.K. were partly supported by The Ministry of Education and Science of the Russian Federation, contract 02.740.11.0250 and RFBR-grants 09-02-12168-ofi_m and 10-02-00249-a.

References

1. F.-K. Thielemann, J. Metzinger and H. V. Klapdor-Kleingrothaus, *Z. Phys. A* **309**, 301 (1983).
2. S. Goriely and B. Clerbaux, *Astron. and Astroph.* **346**, 798 (1999).
3. C. Freiburghaus, J.-F. Rembges, T. T. Rauscher, E. Kolbe, F.-K. Thielemann, K.-L. Kratz, B. Pfeiffer and J. J. Cowan, *ApJ* **516**, 381 (1999).
4. J. Cowan, B. Pfeiffer, K.-L. Kratz, F.-K. Thielemann, C. Sneden, S. Burles, D. Tytler and T. Beers, *ApJ* **521**, 194 (1999).
5. I. V. Panov and F.-K. Thielemann, *Nucl. Phys. A* **718**, 647 (2003).
6. I. Panov and F.-K. Thielemann, *Astronomy Letters* **30**, 647 (2004).
7. G. Martínez-Pinedo, D. Mocelj, N. T. Zinner, A. Kelić, K. Langanke, I. Panov, B. Pfeiffer, T. Rauscher, K.-H. Schmidt and F.-K. Thielemann, *Prog. Part. Nucl. Phys.* **59**, 199 (2007).
8. T. Rauscher and F.-K. Thielemann, *At. Data Nucl. Data Tables* **75**, 1 (2000).
9. Y. Aboussir, J. M. Pearson, A. K. Dutta and F. Tondeur, *At. Data Nucl. Data Tables* **61**, 127 (1995).
10. A. Mamdouh, J. M. Pearson, M. Rayet and F. Tondeur, *Nucl. Phys. A* **664**, 389 (1998).
11. W. D. Myers and W. J. Swiatecki, *Nucl. Phys. A* **601**, 141 (1996).
12. W. D. Myers and W. J. Swiatecki, *Phys. Rev. C* **60**, 014606 (1999).
13. S. Goriely, S. Hilaire, A. J. Koning, M. Sin and C. R., *Phys. Rev. C.* **79**, 024612 (2009).
14. P. Möller, J. R. Nix, W. D. Myers and W. J. Swiatecki, *At. Data Nucl. Data Tables* **59**, 185 (1995).

15. Y. S. Lyutostansky, V. I. Lyashuk and I. V. Panov, *Bull. Acad. Sci. Ussr, Phys. ser.* **54**, 2137 (1990).
16. I. V. Panov, E. Kolbe, B. Pfeiffer, T. Rauscher, K.-L. Kratz and F.-K. Thielemann, *Nucl. Phys. A* **747**, 633 (2005).
17. I. V. Panov, I. Y. Korneev and F.-K. Thielemann, *Physics of Atomic Nuclei* **72**, 1026 (2009).
18. I. V. Panov, I. Korneev, T. Rauscher, G. Martínez-Pinedo, A. Kelić-Heil, N. T. Zinner and F.-K. Thielemann, *Astron. and Astroph.* **513**, A61 (2010).
19. A. S. Krivohatsky and Y. F. Romanov, *Making Transuranium and Actinoid Elements by Neutron Irradiation [in Russian]* (Atomizdat, Moscow, 1970).
20. D. W. Dorn and R. W. Hoff, *Rhys. Rev. Lett.* **14**, 440 (1965).
21. G. I. Bell, *Rhys. Rev.* **139**, B1207 (1965).
22. R. W. Hoff, *J. Phys. G: Nucl. Phys.* **14**, S343 (1988).
23. F.-K. Thielemann, M. Arnould and J. W. Truran, in *Advances in Nuclear Astrophysics*, ed. E. V.-F. et al. (Editions Frontières, Gif-sur-Yvette, 1987).
24. J. J. Cowan, F.-K. Thielemann and J. W. Truran, *Phys. Rep.* **208**, 267 (1991).
25. L. Wolfenstein, *Phys. Rev.* **82**, 690 (1951).
26. W. Hauser and H. Feshbach, *Phys. Rev.* **87**, 366 (1952).
27. F.-K. Thielemann, A. Cameron and J. J. Cowan, in *Fifty Years with Nuclear Fission*, eds. J. Behrens and A. Carlson (American Nuclear Society, Gaithersburg, 1989).
28. S. Bjornholm and J. E. Lynn, *Rev. of Mod. Phys.* **52**, 725 (1980).
29. V. M. Strutinsky, *Nucl. Phys. A* **95**, 420 (1967).
30. P. Möller, J. R. Nix and K.-L. Kratz, *At. Data Nucl. Data Tables* **66**, 131 (1997).
31. T. Rauscher, F.-K. Thielemann and K.-L. Kratz, *Phys. Rev. C* **56**, 1613 (1997).
32. T. Nakagawa, S. Chiba, T. Hayakawa and T. Kajino, *At. Data Nucl. Data Tables* **91**, 77 (2005).
33. P. Möller, A. J. Sierk, T. Ichikawa, A. Iwamoto, R. Bengtsson, H. Uhrenholt and S. Aberg, *Phys. Rev. C* **79**, 064304 (2009).
34. N. Soppera, M. Bossant, H. Henriksson, P. Nagel and Y. Rugama, in *Proc. Int. Conf. on Nuclear Data for Science and Technology, April 22-27, 2007*, ed. e. a. O. Bersillon (EDP Sciences, 2008).
35. F. Tovesson, T. S. Hill, J. D. Baker and C. A. McGrath, *Phys. Rev. C* **79**, 014613 (2009).
36. D. W. Dorn, *Rhys. Rev.* **126**, 693 (1962).
37. I. V. Panov, *Bull. Acad. Sci. Ussr, Phys. ser.* **61**, 210 (1997).
38. K. A. Nedvedjuk and Y. P. Popov, Neutron physics. v. 2, in *Proc. of Workshop on Neutron Physic*, (CNIIAtomInform, Kiev, USSR, 1980).
39. G. I. Bell, *Rev. of Mod. Phys.* **39**, 59 (1967).

HIGH-ENERGY NEUTRON–INDUCED FISSION CROSS SECTION OF SUBACTINIDES

DIEGO TARRÍO, IGNACIO DURÁN, CARLOS PARADELA
Universidade de Santiago de Compostela, Spain

LAURENT TASSAN-GOT, LAURENT AUDOUIN
Centre National de la Recherche Scientifique/IN2P3-IPN Orsay, France

AND THE N_TOF COLLABORATION
CERN, Geneve, Switzerland (www.cern.ch/ntof)

Neutron-induced fission cross sections of ^{nat}Pb and ^{209}Bi between threshold and 1 GeV have been measured at the CERN neutron Time-Of-Flight (n_TOF) facility. The cross section ratios of these subactinides were determined relative to ^{235}U and ^{238}U using PPAC detectors for coincident detection of both fission fragments. There is good agreement with previous experimental results below 200 MeV. Around 1 GeV, the results are compared with the cross sections obtained in proton-induced fission.

1. Introduction

Data on neutron-induced fission cross sections of subactinides are not abundant, mainly due to their low cross sections. However, a good knowledge of this kind of reactions is crucial for the development of Accelerator-Driven Systems (ADS) and of neutron sources in general. In particular, lead and bismuth, which are studied here, will likely be the bulk of the spallation targets [1].

Bismuth is a monoisotopic and non-radioactive material, which makes ^{209}Bi a well suited isotope for reference purposes. The ^{209}Bi(n,f) reaction has been proposed as a cross section standard above 50 MeV. Its high fission threshold (about 20 MeV) makes it a suited fluence monitor of high-energy neutrons, particularly in a large background of low-energy neutrons. Our measurements at CERN-n_TOF of the ^{209}Bi(n,f) and ^{nat}Pb(n,f) cross sections are intended to improve the present evaluations [2] at very high energies.

The high-intensity neutron beam at the n_TOF facility at CERN allowed us to measure very low cross sections. Accordingly, the present work provides the first measurements of the neutron-induced fission cross sections for subactinides covering an extended energy range up to 1 GeV.

2. Experimental facility

The experiment was performed at the neutron Time-Of-Flight (n_TOF) facility at CERN. Spallation neutrons were produced by a pulsed proton beam of 20 GeV/c from the CERN Proton Synchrotron (PS) in a massive lead target. The cooling water surrounding the target acts as a moderator and produces a neutron flux covering a wide energy range, between thermal and several GeV. The high intensity of the neutron beam (about 10^5 neutrons per cm^2 per proton bunch in the experimental area) makes it possible to use radioactive samples. High-resolution energy measurements (with a resolution of 6% around 1 GeV and of 0.01% around 1 eV) are possible due to the long flight path of 185 m between the spallation target and the experimental area [3].

2.1. Fission chamber

Fission events were detected using a dedicated chamber with Parallel Plate Avalanche Counters (PPAC) developed at IPN-Orsay, where samples of ^{235}U and ^{238}U were placed as references. The PPAC detectors consist of one central anode surrounded by two cathodes with a low-pressure gas filling the gaps between the electrodes. Fission fragments are detected in two consecutive PPACs. Because the cathodes of each PPAC detector are segmented in perpendicular directions, the fission fragment trajectory can be reconstructed. The PPAC signals are very fast (about 9 ns FWHM for the anode signals) so that the probability of pile-up is very low, making it possible to reach energies as high as 1 GeV. A total of 10 PPAC detectors and 9 samples in between are placed in the chamber, perpendicularly to the direction of the neutron beam, as shown in Figure 1.

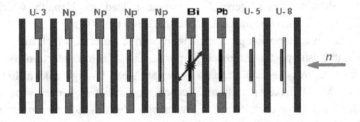

Figure 1. Schematic view of the PPAC detectors and the samples used in this experiment.

Thickness and chemical composition of the samples and of the backings were studied using Rutherford Backscattering Spectroscopy (RBS). For the radioactive samples, a measurement of their α-activity provided an additional measurement of the mass distribution.

3. Data analysis

To reject the background produced by the α-activity of the radioactive samples and by the products of spallation reactions, both fission fragments are detected in two consecutive PPACs within a coincidence window of 10 ns. This is particularly important to identify the fission events above a few hundreds of MeV because in this energy range the probability of producing spallation reactions is around 10 times larger than the probability of inducing a fission event.

Once the fission events are identified, the ratio of the fission cross sections for two samples can be calculated by:

$$\frac{\sigma_i(E)}{\sigma_j(E)} = \frac{C_i(E)}{C_j(E)} \cdot \frac{\Phi_j(E)}{\Phi_i(E)} \cdot \frac{N_j}{N_i} \cdot \frac{\varepsilon_j(E)}{\varepsilon_i(E)} \quad (1)$$

where C(E) is the detected number of fission events, Φ(E) the neutron fluence (in n/cm^2) for the full measuring time, N the total number of atoms in the target, and ε(E) the detection efficiency.

4. Results

Cross sections of ^{nat}Pb(n,f) and ^{209}Bi(n,f) have been measured between threshold and 1 GeV relative to ^{235}U and to ^{238}U. Using the ^{235}U(n,f) and ^{238}U(n,f) reference cross sections from the JENDL/HE-2007 evaluation, the final values of the fission cross-sections ^{nat}Pb(n,f) and ^{209}Bi(n,f) have been obtained. The results for ^{209}Bi(n,f) are shown in Figure 2.

Below 200 MeV our results are in good agreement with previous (n,f) experiments [4, 5]. Above this energy, the evaluated (n,f) cross sections [2, 5], which had to rely on the extrapolation of experimental data, are found to differ significantly from the present results.

Because other experimental (n,f) data above 200 MeV are not available, proton-induced fission cross sections from experimental results [6] and from systematics [7] are shown as well. This comparison indicates that the (n,f) cross section tends to be compatible with the (p,f) data around 1 GeV.

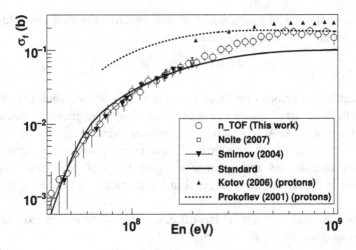

Figure 2. Neutron-induced fission cross section of ^{209}Bi obtained in this work compared with previous data from Refs. [4, 5]. The extension of the present parametrizations [2, 5] above 200 MeV differs significantly from our results. Experimental data [6] and systematics [7] for proton-induced fission are also drawn for comparison.

5. Conclusions

The neutron-induced fission cross sections of ^{nat}Pb and ^{209}Bi have been measured for the first time from threshold up to 1 GeV at the CERN white neutron source n_TOF. The results are in good agreement with previous experiments, which are limited to neutron energies below 200 MeV. Extrapolations of previous data [2,5] to higher energies exhibit significant discrepancies with the new values and can be updated accordingly. Our results are also compared with the available data for protons in the interval from 200 MeV up to 1 GeV.

References

1. OECD-NEA, *Status Report*, NEA-5421 (2005).
2. A. D. Carlson et al., IAEA Report INDC (NDS), 368 (1997).
3. The n_TOF Collaboration, *CERN n_TOF Facility: Performance Report*, CERN/INTC-O-011, CERN-SL-2002-053 ECT (2002).
4. R. Nolte et al, *Nucl. Sci. Eng.* **156**, 197 (2007).
5. A. N. Smirnov et al, *Phys. Rev.* **C70**, 054603 (2004).
6. A. A. Kotov et al, *Phys. Rev.* **C74**, 034605 (2006).
7. A. V. Prokofiev, *Nucl. Inst. And Meth.* **A463**, 557 (2001).

267

PROTON-INDUCED FISSION CROSS SECTION OF ^{181}TA

Y. AYYAD[1], J. BENLLIURE[1], E. CASAREJOS[1], H. ÁLVAREZ-POL[1], A. BACQUIAS[3], A. BOUDARD[3], T. ENQVIST[2], V. FÖHR[2], A. KELIC[2], K. KEZZAR[3], S. LERAY[3], C. PARADELA[1], D. PÉREZ-LOUREIRO[1], R. PLESKAC[2] and D. TARRÍO[1]

[1] *University of Santiago de Compostela, 15754 Santiago de Compostela Spain*

[2] *Helmholtzzentrum für Schwerionenforschung, Planckstrasse 1, 64941 Darmstadt, Germany*

[3] *DSMIRFU/CEA, 91191 Gif-sur-Ivette, France*

We have investigated the total fission cross section of ^{181}Ta+ ^{1}H at 1, 0.8, 0.5 and 0.3 GeV with an specific setup at the FRS (FRagment Separator - GSI). The high-accuracy results obtained in this experiment are compared with calculations performed with an intra-nuclear cascade model (INCL v4.1) coupled to a de-excitation code (ABLAv3p). The calculations reproduce the two different models that describe the fission process at high excitation energies: statistical model of Bohr and Wheeler and the dynamical description of the fission process. Data comparison with previous experiments is also included to point out the existing discrepancies with this new results.

Keywords: tantalum, proton-induced fission, nuclear dissipation

1. Introduction

Fission at medium and high energies provides valuable information on the dynamics of the fission process.[1] Moreover, subactinide data are also relevant for nuclear physics applications such as the construction of spallation neutron sources for Accelerator-Driven Systems (ADS). Despite this recent interest, the available experimental information in this energy domain, and in particular for nuclei with low fissility, is scarce. We have measured the total fission cross-section, at different energies: 300, 500, 800 and 1000 AMeV. By using the inverse kinematic technique and a dedicated experimental setup, we obtained a high-accuracy measurement of the fission channel, which was expected to be about 1% of the total reaction cross-section. This challenging study required a capable and robust experimental setup.

Our results allow for benchmarking the state-of-the-art models, and provide data for systematic descriptions.

2. Experimental Setup

The experiment was performed at FRS (FRagment Separator at GSI - Darmstadt, Germany). The experimental technique was based in inverse kinematics, analyzing the residues of a ^{181}Ta beam. A liquid hydrogen target was surrounded by two Multi Sample Ionization Chambers (MUSIC) where the energy loss allows to identify the reaction products of ^{181}Ta, produced not only in the hydrogen target, but also in the whole target-setup. Downstream the second MUSIC, two scintillator paddles, mounted one on top of the other, provided the fission trigger. The geometry of the trigger fits with the distribution of the fission events that are emitted forward due to the kinematic of the reaction.

3. Data Analysis

In a first step the analysis was performed selecting the data region that corresponds to fission events in the two MUSIC's spectrum (left panel fig. 1). The particles lying in the diagonal of the plot, correspond to ions which keep their atomic number when passing through the target, these are nuclei lighter than the primary beam that have been produced in nuclear reactions induced by ^{181}Ta projectiles in any layer of matter placed upstream the hydrogen target. The ^{181}Ta spot is clearly visible on the top. Here we recognize that many other species arrived to our target, and are eliminated in our analysis to isolate ^{181}Ta induced reactions. The reaction products lay just below the ^{181}Ta spot, in the vertical-aligned group. The evaporation products, immediately below of ^{181}Ta in charge, are the most populated group. Fission products lay in the group of medium charges corresponding to half value of the vertical axis. In a second step, the energy loss in the two scintillator paddles is plot (right panel fig. 1), applying the selection of the fission region observed with the MUSIC detectors. Due to the charge splitting of the fission process, the sum of the energy loss of the resulting residues is constant, and lies in a characteristic region in such a plot. Fission is clearly recognizable in the plot, emerging from the other extremely more populated channels surrounding the region of interest. The selection of this region was made accurately, with the purpose of carefully suppressing the background which remains close to that region.

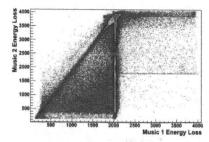 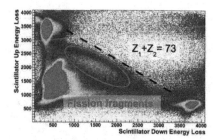

Fig. 1. Left panel: Energy loss in the first MUSIC vs energy loss in the second MUSIC; Right panel : Identification plot provided by the double paddle scintillator. Dashed line corresponds to $Z_1+Z_2=73$. Fission fragments are recognizable inside the selected area.

Fission cross sections were calculated normalizing to the beam dose and target thickness. We also measured the contribution of the target foils (Ti) which surrounded the liquid-hydrogen. This production is subtracted of the total target data, to define the hydrogen contribution.

4. Results and Discussion

We have measured the total fission cross-section of ^{181}Ta+ ^{1}H at different energies, obtaining the values shown in the fig. 2. The results are compared with previous data measured in other experiments and with codes based on a intra-nuclear cascade model (INCL v4.1)[2] which describes the fast interaction between the proton and the Ta, coupled to a de-excitation model (ABLAv3p)[3] describing the later de-excitation of the fragment. These codes are based on the statistical description of the fission rate described by the Bohr-Wheeler transition-state model, and the dynamical description of the fission rate that shows a clear suppression of the fission channel when compared with the prediction of the statistical model.[4] The calculations performed with these codes reveal a difference around one order of magnitude between both models, being the dynamical model, including the dissipation effect, more suitable than the statistical model that largely over-predict the results. We have found a good agreement with the dynamical description of the fission process (compared with the statistical description) that predicts a reduction of the rate due to the introduction of the nuclear dissipation.[5] The results also show that a widely-used systematics[6] over-predict the measured cross sections, in particular, at low energies,diverging from the systematics up to 50% (fig.2).

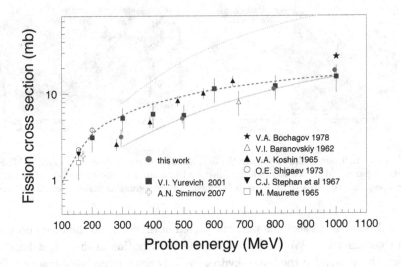

Fig. 2. Total fission cross sections for ^{181}Ta+ ^{1}H. Dotted line: Statistical Model - Solid line: Dynamical model - Dashed line: Prokofiev systematics - Solid circle: This work - Solid Square[7] - Cross[8] - Solid star[9] - Open triangle[10] - Solid triangle[11] - Open circle[12] - Inverted Solid triangle[13] - Open Square[14]

References

1. B. Jurado et al Phys. Rev. Lett. 93(2004) 072501
2. J. Cugnon et al. Nucl. Phys. A 620 (1997) 475
3. J.-J. Gaimard, K-H. Schmidt, Nucl. Phys. A 531 (1991) 709
4. J. Benlliure et al Nucl. Phys. A 700 (2002) 469
5. B. Jurado et al. Phys. Lett. B 553 (2003) 186
6. A.V. Prokofiev Nucl. Inst. Meth. A 463 (2001) 557
7. V.I. Yurevich Fiz. Elementarnykh Chastic i Atomn.Yadra,Letters, Vol.2, p.49 (2005)
8. A.N. Smirnov et al Int Conf on Nuc. Data f Science and Tech (2007)
9. V.A. Bochagov Soviet Journal of Nuclear Physics, Vol.28, p.291 (1978)
10. V.I. Baranovskiy Radiokhimiya, Vol.4, p.470 (1962)
11. V.A. Konshin Yadernaya Fizika, Vol.2, p.682 (1965)
12. O.E. Shigaev Khlopin Radiev. Inst., Leningrad Reports, No.17 (1973)
13. C.J. Stephan Physical Review, Vol.164, p.1528 (1967)
14. M. Maurette Physics and Chemistry of Fission Conf., Salzburg 1965, Vol.2, p.307 (1965)

Fission in the Super-heavy Mass Region

FORMATION AND DECAY OF SUPERHEAVY COMPOUND NUCLEI OBTAINED IN THE REACTIONS WITH HEAVY IONS

M. G. ITKIS, A. A. BOGACHEV, I. M. ITKIS, G. N. KNYAZHEVA, E. M. KOZULIN

*Flerov Laboratory of Nuclear Reactions, Joint Institute for Nuclear Researches
Dubna, 141980, Russia*

> Mass-energy distributions, as well as capture cross-section of fission-like fragments for the reactions of ^{22}Ne, ^{26}Mg, ^{36}S, ^{48}Ca, ^{58}Fe and ^{64}Ni ions with actinides leading to the formation of superheavy compound system with Z=102-120 at energies near the Coulomb barrier have been measured. Fusion-fission cross sections were estimated from the analysis of mass and total kinetic energy distributions. It was found that the fusion probability is approximately the same for the reactions with ^{48}Ca ions and drops three orders of magnitude at the transition to ^{64}Ni ions.

1. Introduction

At present the superheavy elements (SHE) are synthesized in the complete fusion reactions of two heavy nuclei. More than 30 superheavy nuclei (SHE) with Z=108-118 have been synthesized in the reactions of "cold" and "warm" fusion. Whereas in the "cold" fusion reactions the excitation energy of compound nucleus (CN) is 10-20 MeV near the reaction threshold, the CN formation cross section is suppressed strongly by competing quasifission (QF) [1, 2, 3, 4] and deep-inelastic reactions. These processes take place in the "warm" fusion reactions also, but their contributions into the reaction cross sections are less compared with "cold" reactions due to the difference in the entrance channel.

The excitation energy of a formed CN is about 30-40 MeV in "warm" fusion reactions and the de-excitation of CN to the ground state is due to the emission of three or four neutrons and several γ-rays. Nevertheless in the "warm" fusion reactions the formed CN is more neutron rich and nearer to the closed neutron shell at N=184 than in the case of "cold" fusion reactions. Consequently, the "warm" fusion reactions are more preferable for synthesis of the SHE. A big success was achieved in the reactions of actinides with double magic ^{48}Ca ions at FLNR [5]. The production cross sections of SHE in these reactions do not change practically with increasing atomic number of CN and maintain the level of a few picobarn.

One of the possible ways for further progress in the field of the SHE synthesis is to use the complete fusion reactions of ^{238}U, ^{244}Pu and ^{248}Cm nuclei with heavier projectiles, such as ^{58}Fe or ^{64}Ni, leading to the formation of even heavier elements with Z =120–124 and N = 179–183. However, the increase of projectile charge leads to the increasing QF contribution.

This paper presents the results of the experimental investigations of the influence of the entrance channel properties on the competition between fusion-fission (FF) and QF in "warm" fusion reactions. The properties of binary reaction products obtained in the reactions of ^{22}Ne, ^{26}Mg, ^{36}S, ^{48}Ca, ^{58}Fe and ^{64}Ni with actinides have been measured at energies near the Coulomb barrier. Binary reaction products were detected by the two-arm time-of-flight spectrometer CORSET [6].

2. Formation probability of Hs compound nuclei (Z=108) in the reactions with ^{22}Ne, ^{26}Mg, ^{36}S, ^{58}Fe ions

In the reactions chosen the parameter of mass-asymmetry in the entrance channel α varies strongly: for the reaction ^{58}Fe+^{208}Pb: α = 0.571, for ^{36}S+^{238}U: α = 0.737, for ^{26}Mg+^{248}Cm: α = 0.810, and for ^{22}Ne+^{249}Cf: α = 0.838. As demonstrated in Fig. 1 the mass-energy distributions change with decreasing asymmetry α in the entrance channel from symmetric for incoming ^{22}Ne-ions to strongly asymmetric for incoming ^{58}Fe-ions. These changes are understood as reflecting the relative contributions from FF and QF to the fission process of Hs depending on the reaction studied.

It is clearly seen that even at similar CN excitation energies the mass-energy distributions are vastly different for these reactions. In the case of the reactions ^{22}Ne+^{249}Cf and ^{26}Mg+^{248}Cm the mass distributions have a near Gaussian shape with no evidence for asymmetric fission. The reactions are considered to be mainly FF process. For the ^{36}S+^{238}U reaction the mass distributions of the fission-like fragments change markedly. This difference in mass distributions for the ^{26}Mg+^{248}Cm and ^{36}S+^{238}U reactions is connected with an increasing contribution of the QF process for the ^{36}S-induced reaction. At low excitation energies QF is the dominant process for the reaction ^{36}S+^{238}U. At higher excitation energies the mass distribution becomes symmetric and similar to the reaction ^{26}Mg+^{248}Cm though - due to a remaining trace of QF - slightly wider. The obtained mass distributions for the ^{36}S+^{238}U reaction are in good agreement with data from work [7] where only the fragment mass distributions and cross-sections for fission were measured for the same reaction at bombarding energies around the Coulomb barrier.

The mass-energy distribution for the ^{58}Fe+^{208}Pb reaction has the wide two-humped shape even at excitation energy of 48 MeV (well above the Bass barrier). For this reaction the QF process dominates at energies below and above the Bass barrier. This reaction takes a special place due to the fact that one of the partners is double-magic lead. It is known that for heavy-ion induced reaction the formation of QF fragment is determined by the strong influence of the nuclear shell at Z=82 and N=126 (double magic lead). Therefore, for this reaction the QF fragments overlap strongly by quasi-elastic and deep-inelastic events. It is clearly seen on the TKE-mass matrices in the Fig. 1.

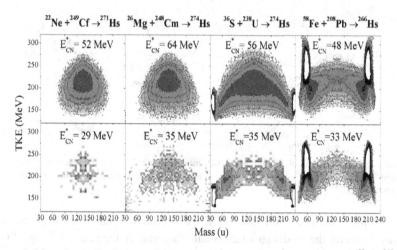

Figure 1. Mass-energy distributions of binary reaction fragments for the reactions ^{22}Ne+^{249}Cf, ^{26}Mg+^{248}Cm, ^{36}S+^{238}U, ^{58}Fe+^{208}Pb leading to the formation of elements with Z=108 at energies above (top panel) and below (bottom panel) the Coulomb barrier.

The QF fragments for the reactions with ^{36}S and ^{58}Fe ions are mainly formed in the mass region near the closed shells with Z=28 and N=50 for the bump at low masses and with Z = 82 and N = 126 for the bump at large masses. The maximum yield corresponds to the fragments with heavy masses 200 u. In the case of the ^{58}Fe+^{208}Pb the separation of quasi-elastic and deep-inelastic events from QF is not so evident in the mass range 200-220 u for heavy fragment. Nevertheless, at the highest energy of ^{58}Fe it is possible to find the maximum of QF component on TKE-mass matrix. It also corresponds to mass ~200 u.

In Fig. 2 the TKE distributions of the fragments in the mass region $A_{CN}/2\pm20$u are presented for these reactions. At excitation energy 35MeV (~10 MeV below the Coulomb barrier for the Mg and S-induced reactions) the TKE distribution of the fragments formed in the reaction ^{26}Mg+^{248}Cm strongly differs from the Gaussian shape and has a high energy component with mean value ~235MeV, while in the reaction ^{36}S+^{238}U the TKE distribution is shifted to the low energy. Moreover, the dispersions of these distributions are different: in the reaction with ^{26}Mg-ions it is 500±20MeV^2, in ^{36}S-induced reaction – 230±20MeV^2.

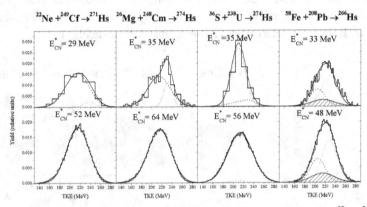

Figure 2. TKE distributions for fragments with masses $A_{CN}/2\pm20$u for the reactions ^{22}Ne+^{249}Cf, ^{26}Mg+^{248}Cm, ^{36}S+^{238}U, ^{58}Fe+^{208}Pb at energies below (top panel) and above (bottom panel) the Coulomb barrier.

At moderate excitation energies (E*=10-30 MeV) shell effects begin to fade away, however the modal structure is still observed. In the reaction ^{26}Mg+^{248}Cm at 35 MeV the high energy component in the TKE distribution may be explained by the phenomenon of bimodal fission. The bimodal fission has been observed for the case of spontaneous and low energy fission of nuclei in the Fm-Rf region [8]. It is important to note that bimodal fission appears when both fission fragments are close to spherical proton (Z=50) and/or neutron (N=82) shells.

So, in the case of the ^{36}S+^{238}U reaction at 35 MeV asymmetric QF gives a big contribution to the symmetric fragment region. It leads to the decrease of the TKE in comparison with the ^{26}Mg+^{248}Cm reaction. A small value of the dispersion for the TKE distribution also confirms the assumption about the QF nature of these fragments. At energy above the Coulomb barrier the TKE distributions are similar for both reactions and well described by Gaussian distribution. Hence, in these cases the symmetric fragments originate from the FF process mainly.

In the case of the ^{58}Fe+^{208}Pb reaction the TKE distributions are practically the same for all measured energies (below and above the Coulomb barrier). So, in the framework of applied analysis it is difficult to conclude about the nature (FF or QF) of symmetric fragments for this reaction.

3. Reactions with ^{48}Ca-ions

Figure 3 displays the obtained distributions of binary fragments obtained in the reaction ^{48}Ca+^{238}U at different excitation energies, namely (from top to bottom): the two-dimensional matrix of events as a function of the mass and total kinetic energy; the mass distribution for fission events framed into the contour line drawn on the two-dimensional mass-energy plot. Reaction products lying between elastic peaks can be identified as totally relaxed events, i.e., as fission (or fission-like) fragments. We have outlined them by solid lines in the panels. Henceforth we consider the properties of these events only.

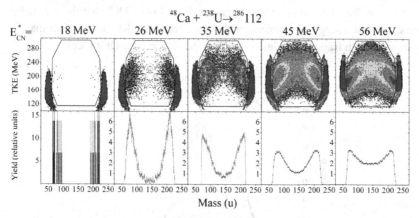

Figure 3. The two-dimensional TKE–M matrices (upper panels) and yields of fragments inside the contour lines on TKE-M matrices (bottom panels) in the 48Ca+238U reaction at projectile energies 212, 222, 232, 244 and 258 MeV corresponding to excitation energies of the CN of 18, 26, 35, 45 and 56 MeV, respectively.

Mass-energy distributions for the reaction ^{48}Ca+^{238}U have the wide two-humped shape caused by QF process mainly determined by the influence of spherical closed shells with Z = 82 and N = 50, 126. The maximum yield corresponds to the fragments with masses about 208 u and complementary light ones.

In spite of domination of asymmetric QF fragments at energies below the coulomb barrier, especially, the yield of symmetric fragments increases with increasing excitation energy.

Figure 4 presents the TKE distribution of fragments with masses $A_{CN}/2\pm20$ u for the reaction ^{48}Ca+^{238}U at energy 232 MeV. It is readily seen that TKE distribution has a complex structure which is not consistent with only FF process. In fact, it is known that in FF process the average TKE of the partner fragments is substantially independent on the excitation energy and shows a typical Gaussian-like shape. The TKE distribution was described by the sum of the three Gaussians with the next values of mean energies and dispersions: 188±3.0 MeV and 213±5 MeV2; 228.5±1.5 MeV and 420±5 MeV2; 265.9±1.9MeV and 74±4 MeV2.

From the Viola systematics [9] we infer that the average TKE is in a first approximation a linear function of the Coulomb barrier $Z^2/A^{1/3}$ whereas from the systematics in Ref. [10] we can estimate the variance of the TKE distribution. For the ^{286}112 CN the variance of the TKE distribution is about 400 MeV^2 and the TKE is 226 MeV. Given the considerable good agreement with the systematics, we associate the middle component of the TKE distribution with the FF process. Since the asymmetric fragments have lower TKE that the symmetric ones, the low energy component of the experimental TKE distribution may be associated with fragments originating from the asymmetric QF process.

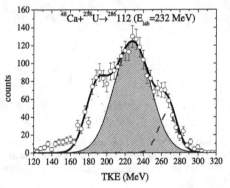

Figure 4. TKE distribution of fragments with masses $A_{CN}/2\pm20$ u for the reaction ^{48}Ca+^{238}U at energy 232 MeV. The open circles are the experimental points, the hatched region corresponds to CN fission with energy taken from the Viola systematic, dashed and dotted curves represent high and low energy components of the TKE distribution.

The high energy part may arise instead from the symmetric mode of the QF. Furthermore, we note that the mean TKE from this mode is about 40 MeV higher that the mean TKE for the FF process. Considering that both processes give rise to symmetric mass fragments, the difference in mean TKE can be taken as an evidence that in the QF process a complete dissipation of the entrance channel energy does not occur. As a consequence, the symmetric fragments with high TKE do not originate from complete fusion because the final fragments retain part of the entrance channel total kinetic energy.

At the transition from ^{238}U to the heavier ^{244}Pu and ^{248}Cm targets the mass-energy distributions of binary reaction products do not change practically. Figure 5 displays the TKE-M matrices for these reactions at CN excitation energy of about 33 MeV. The typical wide two-humped shape with small contribution to the symmetric mass region is observed for these reactions. It is important to note that the relative contribution of the symmetric fragments into the capture cross section does not drop dramatically with increasing of charge of composite system and equal 5-10% at CN excitation of 33 MeV. The evaporation residues cross sections are of the picobarn level for these reactions [5]. Consequently, the reaction mechanism is practically the same for the reactions with the ^{48}Ca ions with actinide targets.

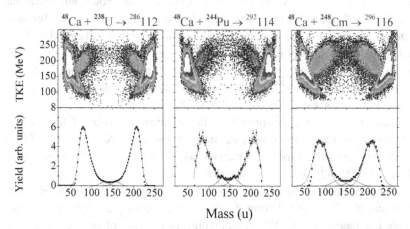

Figure 5. The two-dimensional TKE–M matrices (upper panels) and yields of fission-like fragments (bottom panels) in the ^{48}Ca+^{238}U, ^{244}Pu, ^{248}Cm reactions at the CN excitation energies around 33 MeV.

4. Transition from Ca to Ni ions

Mass-energy distributions of fission fragments have been measured in the ^{44}Ca+^{206}Pb →^{250}No*, ^{64}Ni+^{186}W →^{250}No* at the excitation energies of the CN of 30 and 40 MeV. Figure 6 displays the main characteristics of fission fragment mass-energy distributions for these reactions (from top to bottom: two-dimensional matrix of counts as a function of mass and total kinetic energy; mass distribution for fission events involved into the contour line; average total kinetic energy of fission fragments involved into the contour line as a function of mass).

Mass distributions of fission-like fragments have the complicated structure: the symmetric component typical for the fission of excited CN; the asymmetric fission one connected with the formation of the deformed shell near the heavy fission fragment mass 140 and the asymmetric "shoulders", visible around $Z = 28$ and $N = 50, 88$. In the study of the spontaneous fission properties of heavy actinide nuclei ($Z > 98$) it was found that the transition from asymmetric to symmetric fission in the No isotopes takes place somewhere at $N = 154$ [11], mass distribution of No-isotopes which have less than 154 neutrons is asymmetric and its properties mainly determined by the heavy fragment, peaked around A=140.

In contrast to the reaction with ^{44}Ca, the contribution of the asymmetric "shoulders" into the total mass distribution in the case of ^{64}Ni + ^{186}W greatly increases, the QF is dominating process. The angular momentum for the ^{44}Ca+^{206}Pb and ^{64}Ni+^{186}W systems are similar, so, they should not reveal the significant difference between the mass distributions of CN-fission for both systems. We suggest that the main process for symmetric mass split of the ^{64}Ni+^{186}W system is CN-fission. In order to estimate the upper limit of CN-fission for this reaction, we inscribe the mass distribution for the CN-fission extracted from the ^{44}Ca+^{206}Pb reaction at the same excitation energies in the experimental mass distribution of the ^{64}Ni+^{186}W reaction.

This observation is confirmed by the different behavior of the <TKE> distributions of the fission fragments. In the mass region $A_{CN/2}\pm 20$ the <TKE> distributions are similar for both reactions, while for the asymmetric mass region <TKE> for ^{64}Ni + ^{186}W is higher than that for the ^{44}Ca + ^{206}Pb reaction.

The arrows in Fig. 6 show the positions of the spherical closed shells with Z=28 and N=50, 82 and deformed neutron shell N=88 [12], derived from the simple assumption on the N/Z equilibration. In the case of the ^{44}Ca+^{206}Pb the major part of the QF component fits into the region of these shells, and its maximal yield is a "compromise" between Z=28 and N=50. In the case of the ^{64}Ni+^{186}W the closed shell N=50 and deformed shell with N=88 play important role in the formation of the QF asymmetric component and the drift of mass to the symmetry is more pronounced.

The angular distribution for the asymmetric (where we expected the QF process) and symmetric (where we assume the domination of CN-fission) mass splits for both reactions was measured. In Fig. 7 the angular distributions for the selected mass bins of fission-like fragments are shown. One can see that for both reactions angular distributions are symmetrical for all symmetrical masses, while for the asymmetric mass region the significant forward-backward asymmetry in angular distribution is observed.

From the mass-angle correlation it is evident that the composite system, which suffers the asymmetric mass split (A ≈ 80 u) for both reactions, rotates less than one turn. Consequently, it is possible to estimate the reaction time using these angular distributions in terms of the angle of rotation $\Delta\theta$ of the composite system during the reaction, provided that the relevant angular momentum and moments of inertia are known [2].

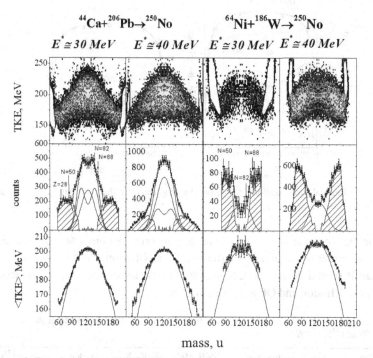

Figure 6. Two-dimensional TKE-mass matrixes (upper panels), yields of fragments and their <TKE> as a function of the fragment mass (middle and bottom panels, respectively) in the $^{44}Ca+^{206}Pb$ (coulombs 1 and 2) and $^{64}Ni+^{186}W$ (coulombs 3, 4) at CN excitation energies 30 and 40 MeV.

For the symmetric mass split the angular distribution is isotropic for the reactions with ^{44}Ca and ^{64}Ni and the value of K_0 agrees well with expected for the CN-fission. It means that typically the nucleus probably rotates several times before scission and the main process, leading to the symmetric mass split, is CN-fission for both reactions. Experimentally, evidence for long CN-fission time scales comes from fission angular distributions, pre-scission neutron measurements also indicate long fission time scale of several 10^{-20} s for CN-fission process [2, 13]. In the paper [2] it was found that the characteristic relaxation time for the mass-asymmetry degree of freedom is $(5.2\pm0.5)\times10^{-21}$s. Besides, it appears that there is time lag of about 2×10^{-21}s before the mass drift starts.

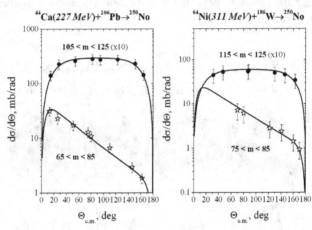

Figure 7. Differential cross section for fission-like fragments for the reactions $^{44}Ca+^{206}Pb$ and $^{64}Ni+^{186}W$ for different mass bins.

For the asymmetric masses where the QF process dominates, the reaction time is one order of magnitude shorter that for CN-fission and the value of K_0 is less than it should be for the CN-fission. The angles of rotations and reaction times for CN-fission and QF are given in Table 1.

Table 1. Angles of rotation and reaction time for CN-fission and QF processes in the reactions $^{44}Ca+^{206}Pb$ and $^{64}Ni+^{186}W$.

Reaction	$<l>$, ℏ	$<m>$, a. u.	TKE, MeV	K_0, ℏ	$\Delta\tau$, deg.	$\Delta\tau$, 10^{-21}s
$^{44}Ca+^{206}Pb$	28	75	177.5	6.0	231	3.6
		116	200	16.7	>270	>32.4
$^{64}Ni+^{186}W$	30	80	187	7.0	226	4.4
		120	200	16.7	>270	>30.2

5. Reactions with ^{58}Fe and ^{64}Ni ions

Unfortunately, the transition to heavier projectile results in the large increase in the Coulomb factors Z_1Z_2 that is crucial in the competition between FF and QF.

The mass-energy distributions for the reactions ^{58}Fe+^{244}Pu and ^{64}Ni+^{238}U (leading to the formation of the same composite system with Z=120 and N=182) at the CN excitation energies about 45 MeV are presented in the figure 8. At first sight the distributions are similar for the both reactions: the wide two-humped shape with large QF component peaked around the mass 215 u. In the formation of the QF component the closed shell at N=50 seems to be effective together with the shell Z=82 and N=126 and leads to the shift of the QF peak from mass 208 u, observed in the reaction ^{48}Ca+^{238}U, to 215 u in the case of the ^{58}Fe+^{244}Pu and ^{64}Ni+^{238}U.

However, at the same CN excitation energy the mass drift to the symmetry (estimated as a distance between masses corresponding to the maximum and half maximum of QF yields) is 22 nucleons in the case of the ^{58}Fe reaction and only 11 nucleons in the case of the ^{64}Ni ions. It is significant that the mass drift to the symmetry is about 34 nucleons for the ^{48}Ca+^{238}U at the same CN excitation energy. The contribution of the symmetric fragments with masses $A_{CN}/2\pm20$ u to all fission-like events is about 8% and 4% for Fe and Ni-ions, respectively. The average TKE's are similar for asymmetric QF fragments for the both reactions, while in the symmetric mass region the TKE for the Fe+Pu reaction is higher than for the Ni+U. The solid line corresponds to the parabolic dependence following from the Liquid Drop Model and the Viola systematics for TKE. The dispersions of the TKE are different for the reactions: it is bigger in the case of Fe-ion induced reaction. In the reaction ^{64}Ni+^{238}U the dispersion does not depend on a fragment mass practically and its mean value is about 350 MeV^2, while for the reaction ^{58}Fe+^{244}Pu dispersion increases for symmetric fragments and its mean value is about 610 MeV^2. According to the existing experimental data on the dispersion of TKE distribution of CN-fission of heavy and superheavy nuclei [10] the dispersion practically does not change for CN with $Z^2/A^{1/3}$ up to ~1000 and increases linearly for heavier CN. The extrapolation of this experimental dependence gives the value of about 550 MeV^2 for the fission of ^{302}120. Hence, for the reaction with Ni-ions TKE dispersion is lower than this value, whereas it is higher in the case of Fe-ions.

In the figure 9 the TKE distributions of fission-like fragments in the mass region $A_{CN}/2\pm20$ u for the reactions ^{58}Fe+^{244}Pu and ^{64}Ni+^{238}U are presented. It is seen that both TKE distributions have a complex structure as in the case of the ^{48}Ca+^{238}U reaction.

In contrast to the ^{58}Fe+^{244}Pu, for the reaction ^{64}Ni+^{238}U the TKE distribution has more pronounced low and high energy components (see fig. 9b), while the component with average value of 252 MeV (corresponding to the Viola systematics) is highly hindered. Because of the low statistics, only an upper value for the relative yield of the CN-fission component can be reasonably given. Table 2 gives the relative contribution of the all symmetric fragments in the mass range $A_{CN}/2 \pm 20$ u and symmetric fragments with TKE corresponding to the Viola systematics.

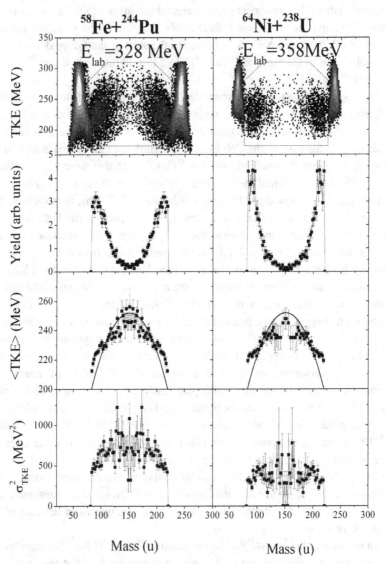

Figure 8. From top to bottom: the two-dimensional TKE–M matrices, yields, average TKEs and TKE dispersions of fragments inside the contour lines on TKE-M matrices (bottom panels) in the ^{58}Fe+^{244}Pu and ^{64}Ni+^{238}U reactions at the CN excitation energies about 45 MeV.

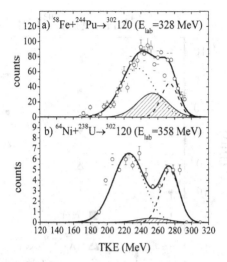

Figure 9. TKE distributions of fragments with masses $A_{CN}/2\pm20$ u. for the reactions $^{58}Fe+^{244}Pu$ and $^{64}Ni+^{238}U$ reaction at the CN excitation energies about 45 MeV.

The obtained captures cross sections as well as cross section for formation of symmetric fragments with mass $A_{CN}/2\pm20$ u are presented in the figure 10 for the reaction $^{64}Ni + ^{238}U$. The capture cross sections are about a few hundred millibarns for Ca and Ni induced reactions, whereas the formation of symmetric fragments is one order of magnitude less for the reaction $^{64}Ni + ^{238}U$. Yet, in the case of the Ca+U at the highest energy, approximately 70% of the events have the TKE expected for the CN fission process, whereas in the case of the Ni reaction only a few percent of symmetric fragments have the TKE compatible with the Viola prediction for the $^{302}120$ CN-fission. While the $^{64}Ni + ^{238}U$ reaction has lower excitation energy at the reaction threshold energy, the FF cross section is suppressed by stronger QF process.

Table 2. The relative contributions of all symmetric fragments ($\sigma_{ACN/2\pm20}$) and symmetric fragments with TKE corresponding to the older Viola systematics (σ_{FF}) to the capture cross section (σ_{cap}) for the reactions $^{48}Ca+^{238}U$, $^{58}Fe+^{244}Pu$ and $^{64}Ni+^{238}U$ at CN excitation energy of around 45 MeV.

Reaction	$\sigma_{ACN/2\pm20}/\sigma_{cap}$ (%)	$\sigma_{FF}/\sigma_{ACN/2\pm20}$ (%)	σ_{FF}/σ_{cap} (%)
$^{48}Ca+^{238}U$	12±2	68±3	8±4
$^{58}Fe+^{244}Pu$	8±3	≤25	≤2
$^{64}Ni+^{238}U$	4±1	≤5	≤0.2

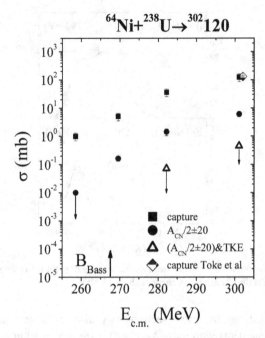

Figure 10. Capture cross section (squares), cross section for the formation of fragments with masses $A_{CN}/2\pm20$ (circles) and the fragments, but with TKE corresponding to the Viola systematic (open triangles) for the ^{64}Ni+^{238}U reaction. Rhomb is the capture cross section from [2].

6. Summary

Mass and energy distributions of fragments, fission and quasifission cross sections, have been studied for a wide range of nuclei with $Z = 102\text{-}122$ produced in reactions with ^{22}Ne, ^{26}Mg, ^{36}S, ^{48}Ca, ^{58}Fe and ^{64}Ni ions at energies close and below the Coulomb barrier.

For the region of very heavy nuclei with compound masses from A≈256 to A≈280 the fragment mass distributions are in most cases symmetric. When both fragments are close to either Z=50 or N=82 bimodal fission shows up. The bimodal fission caused by clustering phenomena was observed for fission of superheavy nuclei ^{271,274}Hs. In the case of fission of superheavy nuclei the light fragments with mass 132-134 u play the stabilizing role.

The major part of the quasifission fragments peaks around the region of the Z=82 and N=126 (double magic lead) and N=50 shells, and the maximum of the yield of the quasifission component is a mixing between all these shells. Hence, shell effects are present everywhere and determine the basic characteristics of fragment mass distributions.

At the transition from Ca to Ni projectiles the contribution of quasifission process rises sharply and Ni ions is not suitable for the synthesis of element Z=120 in the complete fusion reactions.

References

1. R. Bock et al., *Nucl. Phys.* **A388** 334 (1982).
2. Töke et al., *Nucl. Phys.* **A440** 327 (1985).
3. W. Q. Shen et al., *Phys. Rev.* **C 36** 115 (1987).
4. M. G. Itkis et al., *Inter. Jour. Mod. Phys.* **E 16** 957 (2007).
5. Yu. Ts. Oganessian, J. Phys. G: *Nucl. Part. Phys.* **34** R165–R242 (2007).
6. E. M. Kozulin et al., *Instr. Exp. Tech.* **51** 44 (2008).
7. K. Nishio et al., *Phys. Rev.* **C 77** 064607 (2008).
8. E. K. Hulet et al., *Phys. Rev.* **C 40** 770 (1989).
9. V. E. Viola, *Nucl. Data Tables* **A1** 391 (1966).
10. M. G. Itkis and A. Ya. Rusanov, *Fiz. Elem. Chastits At. Yadra* **29**, 389 (1998) [*Phys. Part. Nucl.* **29**, 160 (1998)].
11. E. K. Hulet, *Yad. Fiz.* **57**, 1165 (1994).
12. B. D. Wilkins, E. P. Steiberg and R. R. Chasman, *Phys. Rev.* **C14** 1832 (1976).
13. D. J. Hinde et al., *Nucl. Phys.* **A452** 550 (1986).

… 289

TERNARY FISSION AND QUASI-FISSION OF SUPERHEAVY NUCLEI AND GIANT NUCLEAR SYSTEMS

V.I. ZAGREBAEV, A.V. KARPOV

Flerov Laboratory of Nuclear Reactions, JINR, Dubna, Moscow Region, Russia

WALTER GREINER

Frankfurt Institute for Advanced Studies, J.W. Goethe-Universität, Germany

We found that a true ternary fission with formation of a heavy third fragment (a new kind of radioactivity) is quite possible for superheavy nuclei due to the strong shell effects leading to a three-body clusterization with the two doubly magic tin-like cores. The three-body quasi-fission process could be even more pronounced for giant nuclear systems formed in collisions of heavy actinide nuclei. In this case a three-body clusterization might be proved experimentally by detection of two coincident lead-like fragments in low-energy U+U collisions.

1. Introduction

Today the term "ternary fission" is commonly used to denote the process of formation of light charged particle accompanied fission [1]. This is a rare process (less than 1%) relative to binary fission, see Fig. 1. As can be seen the probability of such a process decreases sharply with increasing mass number of the accompanied third particle. These light particles are emitted almost perpendicularly with respect to the fission axis (equatorial emission) [1]. It is interpreted as an indication that the light ternary particles are emitted from the neck region and are accelerated by the Coulomb fields of both heavy fragments.

In contrast to such a process, the term "true ternary fission" is used for a simultaneous decay of a heavy nucleus into three fragments of not very different mass [1]. Such decays of low excited heavy nuclei were not unambiguously observed yet. The true ternary fission of atomic nuclei (below we omit the word "true") has a long history of theoretical and experimental studies. Early theoretical considerations based on the liquid drop model (LDM) [3] showed that for heavy nuclei ternary fission produces a larger total energy release in comparison to binary fission, but the actual possibility of ternary fission is decided, in fact, by barrier properties and not by the total energy release. It was found that the LDM ternary fission barriers for oblate (triangle) deformations are much higher as compared to the barriers of prolate configurations [4], and it seems that the oblate ternary fission may be excluded from consideration. However further study of this problem within the more sophisticated three-center shell model [5] showed that the shell effects may significantly reduce the ternary fission barriers even for oblate deformations of very heavy nuclei.

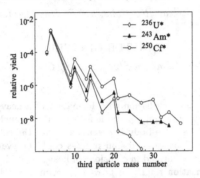

Figure 1. Relative to binary fission yields of ternary particles in the (n_{th}, f) reactions. The figure is a simplified version of Fig. 4 from [2] kindly prepared by F. Gönnenwein.

It is well known that for superheavy nuclei the LDM fission barriers are rather low (or vanish completely) and the shell correction to the total deformation energy is very important. First estimations of the binary and prolate ternary fission barriers of superheavy nucleus $^{298}114$, made in [6] with the shell corrections calculated in an approximate way, demonstrated that they are identical to within 10%. Dynamical aspects of the true ternary fission of very heavy nuclear systems (treated as a neck instability within the LDM) and its dependence on nuclear viscosity were discussed in [7]. To our knowledge, since then there was not any significant progress in theoretical (or experimental) study of ternary fission. In the meanwhile, today it becomes possible to study experimentally the properties and dynamics of formation and decay of superheavy nuclei [8], for which the ternary fission could be rather probable (see below).

2. Clusterization and shape isomeric states of heavy nuclei

The two-center shell model (TCSM) [9] looks most appropriate for calculation of the adiabatic potential energy of heavy nucleus at large dynamic deformations up to the configuration of two separated fragments. The nuclear shape in this model is determined by 5 parameters: the elongation R of the system, which for separated nuclei is the distance between their mass centers; the ellipsoidal deformations of the two parts of the system δ_1 and δ_2; the mass-asymmetry parameter $\eta = (A_2 - A_1)/(A_2 + A_1)$, where A_1 and A_2 are the mass numbers of the system halves; and the neck parameter ϵ which smoothes the shape of overlapping nuclei.

Within the macro-microscopic approaches the energy of the deformed nucleus is composed of the two parts $E(A, Z; R, \delta, \eta, \epsilon) = E_{\text{mac}}(A, Z; R, \delta, \eta, \epsilon) + \delta E(A, Z; R, \delta, \eta, \epsilon)$. The macroscopic part, E_{mac}, smoothly depends on the proton and neutron numbers and may be calculated within the LDM. The microscopic part, δE, describes the shell effects. It is constructed from the single-particle energy spectra by the Strutinsky procedure [10]. The details of calculation of the single particle energy spectra within the TCSM, the explanation of all the parameters used as well as the extended and empirical versions of the TCSM may be found in [11].

Within the TCSM for a given nuclear configuration $(R, \eta, \delta_1, \delta_2)$ we may unambiguously determine the two deformed cores a_1 and a_2 surrounded with a certain number of shared nucleons $\Delta A = A_{\text{CN}} - a_1 - a_2$ (see Fig. 2). During binary fission these valence nucleons gradually spread between the two cores with formation of two final fragments A_1 and A_2. Thus, the processes of compound nucleus (CN) formation, binary fission and quasi-fission may be described both in the space of the shape parameters $(R, \eta, \delta_1, \delta_2)$ and in the space $(a_1, \delta_1, a_2, \delta_2)$. This double choice of equivalent sets of coordinates is extremely important for a clear understanding and interpretation of the physical meaning of the intermediate local minima appearing on the multi-dimensional adiabatic potential energy surface and could be used for extension of the model for description of three-core configurations appearing in ternary fission.

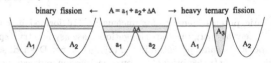

Figure 2. Schematic view of binary and ternary fission.

The adiabatic driving potential for formation and decay of the superheavy nucleus $^{296}116$ at fixed deformations of both fragments is shown in Fig. 3 as a function of elongation and mass asymmetry and also as a function of charge numbers z_1 and z_2 of the two cores (minimized over neutron numbers n_1 and n_2) at $R \leq R_{\text{cont}}$. Following the fission path (dotted curves in Fig. 3a,b) the nuclear system passes through the optimal configurations (with minimal potential energy) and overcomes the multi-humped fission barrier. The intermediate minima located along this path correspond to the shape isomeric states. These isomeric states are nothing else but the two-cluster configurations with magic or semi-magic cores surrounded with a certain amount of shared nucleons. In the case of binary fission of nucleus $^{296}116$ the second (after ground state) minimum on the fission path arises from the two-cluster nuclear configuration consisting of tin-like ($z_1 = 50$) and krypton-like ($z_2 = 36$) cores and about 70 shared nucleons. The third minimum corresponds to the mass-symmetric clusterization with two magic tin cores surrounded with about 30 common nucleons.

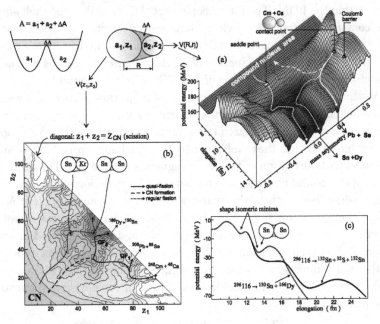

Figure 3. Adiabatic potential energy for nucleus $^{296}116$ formed in collision of ^{48}Ca with ^{248}Cm. (a) Potential energy in the "elongation-mass asymmetry" space. (b) Topographical landscape of the same potential in the (z_1, z_2) plane. Dashed, solid and dotted curves show most probable trajectories of fusion, quasi-fission and regular fission, respectively. The diagonal corresponds to the contact configurations ($R = R_{cont}, z_1 + z_2 = Z_{CN}, \Delta A = 0$). (c) Potential energy calculated for binary (dotted curve) and symmetric ternary fission of nucleus $^{296}116$ (see below).

A three-body clusterization might appear just on the path from the saddle point to scission, where the shared nucleons ΔA may form a third fragment located between the two heavy clusters a_1 and a_2. In Fig. 2 a schematic view is shown for binary and ternary fission starting from the configuration of the last shape isomeric minimum of CN consisting of two magic tin cores and about 30 extra (valence) nucleons shared between the two clusters and moving initially in the whole volume of the mono-nucleus. In the case of two-body fission of $^{296}116$ nucleus these extra nucleons gradually pass into one of the fragments with formation of two nuclei in the exit channel (Sn and Dy in our case, see the fission path in Fig. 3). However there is a chance for these extra nucleons ΔA to concentrate in the neck region between the two cores and form finally the third fission fragment.

3. True ternary fission of superheavy nuclei

There are too many collective degrees of freedom needed for proper description of the potential energy of a nuclear configuration consisting of three deformed heavy fragments. We restricted ourselves by consideration of the potential energy of a three-body symmetric configuration with two equal cores $a_1 = a_2$ (and, thus, with two equal fragments $A_1 = A_2$ in the exit fission channels). Also we assume equal dynamic deformations of all the fragments, $\delta_1 = \delta_2 = \delta_3 = \delta$, and use the same shape parametrization for axially symmetric ternary fission as in Ref. [12] (determined by three smoothed oscillator potentials).

The third fragment, a_3, appears between the two cores when the total elongation of the system, described by the variable R (distance between a_1 and a_2), is sufficiently large to contain all three fragments, i.e., $R \geq R(a_1) + 2R(a_3) + R(a_2)$. Finally, we calculated the three-dimensional potential energy $V(R, \delta, A_3)$ trying to find a preferable path for ternary fission and estimate how much larger the barrier is for three-body decay as compared to binary fission. For better visualization we plot the calculated potential energy $V(R, \delta, A_3)$ as a function of $(R/R_0 - 1)\cos(\alpha_3)$ and $(R/R_0 - 1)\sin(\alpha_3)$ at fixed dynamic deformation $\delta = 0.2$, where $\alpha_3 = \pi \cdot A_3/100$ and R_0 is the radius of sphere of equivalent volume (CN).

The macroscopic (LDM) part of the potential energy for ^{248}Cm is shown in upper-left panel of Fig. 4. The binary fission of ^{248}Cm evidently dominates because after the barrier the potential energy is much steeper just in the binary exit channel (right bottom corner, $A_3 \sim 0$). Emission of light third particle is possible here but not the true ternary fission. The shell correction (which makes deeper the ground state of this nucleus by about 3 MeV) does not change distinctively the total potential energy (see the upper-right panel of Fig. 4). The reason for that is quite simple. For nuclei with $Z < 100$ there is just not enough charge and mass to form two doubly magic tin-like nuclei plus a third heavy fragment. Nevertheless the experiments aimed on the observation of real ternary fission of actinide nuclei (with formation of heavy third fragment) are currently in progress [13].

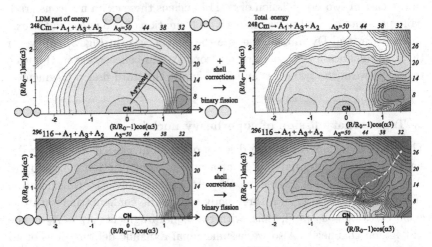

Figure 4. Potential energy for ternary fission of ^{248}Cm (upper plots) and superheavy nucleus ^{296}116 (bottom). Macroscopic part of potential energy and the total one (LDM plus shell corrections) are shown at the left and right panels, respectively, depending on elongation and mass of third fragment (italic numbers). The dashed curve shows the most probable ternary fission of nucleus ^{296}116 onto ^{132}Sn+^{32}S+^{132}Sn.

In the case of superheavy nuclei the macroscopic potential energy does not lead to any barrier at all (neither in binary nor in ternary exit channel) and stability of these nuclei is determined completely by the shell corrections. In bottom panels of Fig. 4 the calculated potential energy is shown for superheavy nucleus ^{296}116. In contrast with ^{248}Cm, in this case a real possibility for ternary fission appears with formation of third fragment $A_3 \sim 30$ and two heavy fragments $A_1 = A_2 \sim 130$. The ternary fission valley is quite well separated by the potential ridge from the binary fission valley. This means that the ternary fission of ^{296}116 nucleus into the "tin–sulfur–tin" combination should dominate as compared with other true ternary fission channels of this nucleus.

More sophisticated consideration of the multi-dimensional potential energy surface is needed to estimate the "ternary fission barrier" accurately. However, as can be seen from Fig. 4, the height of the ternary fission barrier is not immensely high. It is quite comparable with the regular fission barrier because the ternary fission starts in fact from the configuration of the shape isomeric state which is located outside from the first (highest) saddle point of superheavy nucleus $^{296}116$ [see the solid curve in Fig. 3(c)].

4. Ternary quasi-fission of giant nuclear systems

Similar process of decay onto three doubly magic heavy fragments might occur also for giant nuclear systems formed in low-energy collisions of actinide nuclei. In this case compound nucleus hardly may be formed, and such decay is, in fact, a quasi-fission process. Conditions for the three-body decay are even better here, because the shell effects significantly reduce the potential energy of the three-cluster configurations with two strongly bound lead-like fragments. In Fig. 5 the landscape of the potential energy surface is shown for a three-body clusterization of the nuclear system formed in collision of U+U.

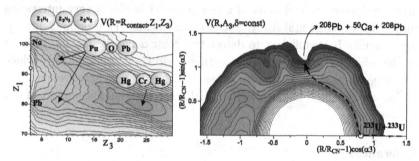

Figure 5. (Left panel) Landscape of potential energy of three-body contact configurations of giant nuclear system formed in collision of ^{238}U+^{238}U. (Right panel) The same as in Fig. 4 but for ^{233}U+^{233}U collision.

In the left panel the potential energy is shown as a function of three variables, Z_1, Z_3 and R (minimized over the neutron numbers) at fixed (equal) deformations of the fragments being in contact $(R_1 + 2R_3 + R_2 = R)$. As can be seen, the giant nuclear system, consisting of two touching uranium nuclei, may split into the two-body exit channel with formation of lead-like fragment and complementary superheavy nucleus (the so-called antisymmetrizing quasi-fission process which may lead to an enhanced yield of SH nuclei in multi-nucleon transfer reactions [14]). Beside the two-body Pb–No clusterization and the shallow local three-body minimum with formation of light intermediate oxygen-like cluster, the potential energy has the very deep minimum corresponding to the Pb-Ca-Pb-like configuration (or Hg-Cr-Hg) caused by the N=126 and Z=82 nuclear shells. In the right panel of Fig. 5 the potential energy of the giant nuclear system formed in collision of ^{233}U+^{233}U is shown as a function of $(R/R_0 - 1)\cos(\alpha_3)$ and

$(R/R_0 - 1)\sin(\alpha_3)$ (see above). A possible ternary decay of this system into ^{208}Pb+^{50}Ca+^{208}Pb is shown by the dashed curve.

5. Summary

Thus we found that for superheavy nuclei the three-body clusterization (and, hence, real ternary fission with a heavy third fragment) is quite possible. The simplest way to discover this phenomenon is a detection of two tin or xenon-like clusters in low energy collisions of medium mass nuclei with actinide targets, for example, in ^{64}Ni+^{238}U reaction. These unusual decays could be searched for also among the spontaneous fission events of superheavy nuclei [8].

The extreme clustering process of formation of two lead-like doubly magic fragments in collisions of actinide nuclei is also a very interesting subject for experimental study. Such measurements, in our opinion, are not too difficult. It is sufficient to detect two coincident lead-like ejectiles (or one lead-like and one calcium-like fragments) in U+U collisions to conclude unambiguously about the ternary fission of the giant nuclear system.

We are indebted to the DFG – RFBR collaboration for support of our studies.

References

1. C. Wagemans, *Ternary Fission* in *The Nuclear Fission Process*, edited by Cyriel Wagemans (CRC Press, Boca Raton, 1991), Chap. 12.
2. F. Gönnenwein et al., in *Seminar on Fission*, Pont d'Oye IV, Belgium, 1999, edited by Cyriel Wagemans et al., (World Scientific, Singapore, 1999) p. 59.
3. W.J. Swiatecki, Second UN Int. Conf. on the Peaceful Uses of Atomic Energy, Geneva, 1958, p. 651.
4. H. Diehl and W. Greiner, Nucl. Phys. **A229**, 29 (1974).
5. A.R. Degheidy, J.A. Maruhn, Z. Phys. A **290**, 205 (1979).
6. H. Schulheis, R. Schulheis, Phys. Lett. B **49**, 423 (1974).
7. N. Carjan, A.J. Sierk and J.R. Nix, Nucl. Phys. **A452**, 381 (1986).
8. Yuri Oganessian, J. Phys. G, **34**, R165 (2007).
9. J. Maruhn and W. Greiner, Z. Phys. A **251**, 431 (1972).
10. V.M. Strutinsky, Nucl. Phys. A **95**, 420 (1967); Nucl. Phys. A **122**, 1 (1968).
11. V.I. Zagrebaev, A.V. Karpov, Y. Aritomo, M. Naumenko, W. Greiner, Phys. Part. Nucl. **38**, 469 (2007).
12. X. Wu, J. Maruhn and W. Greiner, J. Phys. G **10**, 645 (1984).
13. D.V. Kamanin, Yu.V. Pyatkov, A.N. Tyukavkin and Yu.N. Kopatch, Int. J. Mod. Phys. E**17**, 2250 (2008).
14. V.I. Zagrebaev, Yu.Ts. Oganessian, M.G. Itkis and W. Greiner, Phys. Rev. C **73**, 031602 (2006).

WHAT CAN WE LEARN FROM THE FISSION OF THE SUPER-HEAVY ELEMENTS?

DAVID BOILLEY*

GANIL, CNRS/IN2P3-CEA/DSM, BP 55027, 14076 Caen cedex 5, France, and Univ. Caen, Esplanade de la Paix, B.P. 5186, 14032 Caen cedex, France
E-mail: boilley@ganil.fr

YOANN LALLOUET

IPN Lyon, CNRS/IN2P3, Université Lyon1, 69622 Villeurbanne Cedex, France

BÜLENT YILMAZ

Department of Physics, Ankara University, 06100 Ankara, Turkey

YASUHISA ABE

RCNP, Osaka University, Ibaraki (Osaka), 567-0047 Japan

Nuclear shell model calculations predict the existence of super-heavy elements (SHE) that are tentatively synthesized through heavy-ion collisions. A complete description of the reaction to synthesize super-heavy elements is necessary to bridge these predictions with the experimental results on the fission time and residue cross sections. In this contribution, we will present the constraints that can be given on the shell correction energy from experimental data and the developments that are needed for the dynamical models. We will especially focus on the fission time of heavy elements and on the role of the isomeric potential pockets.

Keywords: Super-Heavy Elements; Fission time; Isomeric state.

1. Introduction

The size of the nuclei in nature is limited. But super-heavy elements are expected to exist beyond uranium due to an extra-stability given by the next shell closure for the nucleons. There has been a long quest to synthesize these elements by heavy ion collisions in various laboratories. Experimentally, the main difficulties arise from the fact that such a reaction is not favourable and the cross sections are extremely small, of the order of few picobarns, or even less.

It is very important to note that these elements should not exist if one only considers the Liquid Drop Model. Therefore, these are very fragile objects that easily decay through fission as soon as they are slightly excited. Their main properties come from the shell structure, but there are still many ambiguities on the Z of the next shell closure and on the absolute value of the shell correction energy. At GANIL there is also a tentative to locate the super-heavy island of stability by measuring the fission time. Recent experiments based on crystal blocking techniques have shown that the $Z = 120$ and 124 elements have a long fission time, suggesting an extra-stability.

There is a need for theoretical developments on the description of the whole reaction processes between the two colliding nuclei up to the super-heavy element. The heavier elements formed up to now where identified by their alpha-decay chain and their properties are unknown. Then, a well understanding of the reaction mechanism is also necessary to link the shell correction energy predicted by structure models to the experimental results.

Actually, the fusion mechanism is not a simple extrapolation of what is known with lighter nuclei. It is well known that fusion is hindered in this region, i.e. the fusion cross section is far lower than one would expect. The origin of the fusion hindrance is nowadays well understood on a qualitative point of view,[1] but they are still many quantitative ambiguities. Therefore, we have not reached yet the state of being able to guide the experiments without ambiguity.

2. Residue cross sections

Super-heavy nuclei mainly decay though fission, but we are interested in the small neutron-evaporation channel that stabilizes the nucleus. In order to calculate this very tiny fraction, we have developed a fission-evaporation code that can calculate very low cross sections in a short time.

2.1. The Kewpie2 code

The Kewpie2 code[2,3] is based on the Bateman equations describing the time evolution of an evaporation cascade, including neutrons, protons, alphas, gammas... The physical ingredients are the usual ones: it can accommodate both Weisskopf and Hauser-Feschbach evaporation widths. The fission width is based on the Bohr and Wheeler formula with Kramers and Struntinsky correction factors. The collective enhancement factor is also included. For details and references, see Ref. 3.

The level density parameter is taken from Töke-Swiatecki and the damping of the shell correction energy with the excitation energy follows Ignatyuk's prescription: at the ground state, the level density parameter reads

$$a_g = a\left[1 + \frac{(1 - e^{-E^*/E_d})\Delta E_{shell}}{E^*}\right], \quad (1)$$

where the damping energy is set to its usual value, $E_d = 18.5$ MeV.

The main particularity of this code is that it is not based on a Monte-Carlo algorithm that is not well suited for very low probabilities.

2.2. Evaporation residues

The fission channel dominates the disintegration of the compound nucleus formed by heavy-ions collision. If we tune slightly the fission width, this will not affect much the fission probability that remains close to 1, but it will dramatically change the fate of the evaporation residue. The fission width mainly depends on three parameters that are the fission barrier height that mainly consists of the shell correction energy, the damping energy E_d, see Eq. (1) and the reduced friction parameter. If these last two parameters are fixed to their usual values, the measured residue cross section can constrain the shell correction energy with a precision of 1 MeV. This accuracy corresponds to about one order of magnitude in the residue cross section.

Unfortunately, such a precision can only be obtained if we know precisely the fusion probability. But it is well known that the fusion mechanism is hindered for heavy elements because of the appearance of the so-called quasi-fission process. Experimentally, it is very difficult to distinguish between fission and quasi-fission, and then to evaluate the fusion probability without ambiguity. On a theoretical point of view, it is commonly accepted that the fusion hindrance is due to the appearance of an additional inner barrier that has to be crossed after the Coulomb barrier, but the various models differ on the size of this barrier and on the strength of the dissipation mechanism. Therefore, the main challenge is to find ways to assess the fusion models by other means.[1,4]

One of the ways to get rid of these problems is to send the projectile at energy well above the barriers in order to have a large fusion probability. Then, the compound nucleus will have no chance to survive, but one can get some information by measuring its fission time. This is the topic of the second part of this presentation.

3. Fission time measurements

The fission time of the $Z = 114$, 120 and 124 nuclei was measured at GANIL using the crystal blocking technique.[5] It has been found that for the $Z = 120$ and 124 nuclei, at least 10% of the capture events had a fission time longer than 10^{-18} s, which is very long. No such events were observed for the $Z = 114$ nucleus.

Such a long fission time cannot be calculated using a Langevin equation, as it is traditionaly done.[6] But the Kewpie2 code that solves Bateman equation in time can calculate dynamical observables.[3,7] It appears that whatever the mass table we use as an input of the code, we cannot reproduce such a statistics for the fission times longer than 10^{-18} s.

How can we understand such results? Some hints will be given, using a simplified model.

With excitation energy of the order of 70 MeV, we can safely neglect the evaporation of charged particles like protons and alphas. We will therefore only consider neutrons and gammas. The characteristics of the nuclei entering the evaporation chain are not known. As a toy, model, we will first fix the fission barrier of each isotope of the chain to an identical value, B_f.

The average fission time is plotted as a function of the fission barrier compared to the neutron binding energy B_n in Fig. 1.

With a small fission barrier, fission occurs rapidly at the beginning of the chain. When B_f increases, the fission time increases. But for large barriers, it is the opposite. In this case, fission events are becoming rare and are mainly first chance fission. After the evaporation of few neutrons, the nucleus is too cold to undergo fission.

This means that the long fission times that were observed correspond to fission events that occurred after evaporating several neutrons. Then, in order to reproduce the experimental data, one has to guess the fission barrier or shell correction energy of several isotopes, up to 9. The only thing we can say is that large shell correction energies are necessary to reproduce the data, far larger than the prediction of any mass table.

In this model, the description of the fission width is based on the Bohr and Wheeler model with a single saddle. But there are predictions[8] that the potential landscape along the fission path has several humps. How does it influence the fission time?

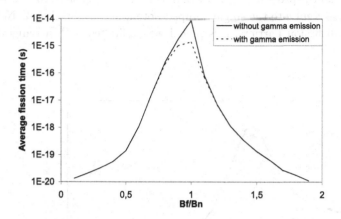

Fig. 1. Average fission time as a function of B_f/B_n assumed to be the same for all the isotopes of a $Z = 120$ like nucleus at an excitation energy of 70 MeV.

4. Influence of structures in the potential on the fission time[10]

It is well known that in the actinides region, the potential has a complex structure along the fission pathway. It might be the same in the super-heavy region. Then, we cannot simply apply the previous model based on a single saddle.

There are various theoretical tools in the literature to evaluate the average fission time.[9] Solving numerically the Langevin equation or using the so-called Non Linear Relaxation Time formula, we can show the largest effect on the average fission time with a double-humped potential is when the barriers have the same size (see the dashed curve of Fig. 3). Then, the average fission time is three times longer than with a single barrier having the same height. We will assume such a potential in the following.

The Langevin formalism including neutron evaporation that is usually used to calculate the fission dynamics[6] can hardly be applied in this context because of the extremely long fission times we need to calculate. We have developed another model based on master equations.

To estimate the rate of jumping into the other potential well or to escape, we use Kramers formula. Evaporation of neutrons that cools down the nuclei is estimated within the Weisskopf formalism. Assuming that the potential structure is the same for all isotopes, we can calculate the average fission time and the probability to have a fission time longer than 10^{-18}s and compare these results to the single humped potential case. See Fig. 2. Note that for small fission barriers, this model was validated with a Langevin type approach.

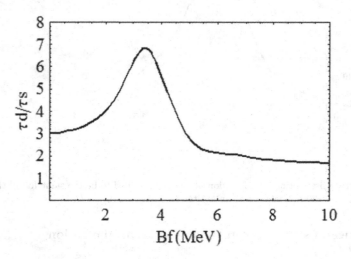

Fig. 2. Average fission time calculated with a double-humped potential divided by the same time calculated with a single-humped potential as a function of the fission barrier B_f. B_n is fixed to 6 MeV.

It can be noted that structures in the potential can naturally enlarge the average fission time of at most a factor 7.

Of course the assumption of a uniform potential for all the isotopes is not realistic. It should not be the same for each isotope, and especially structure should disappear at high excitation energy.[8] In order to evaluate this effect, we have considered a potential depending on the excitation energy as shown on Fig. 3 and solved numerically the Langevin equation with neutron evaporation. We have found that there is almost no difference on the average fission time and the number of events having a fission time longer than 10^{-18} s between a single-humped potential and a double-humped potential.

This means that the large tail of the long fission time distribution that was observed in the experiment cannot be explained by the structure of the potential.

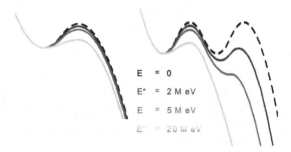

Fig. 3. Evolution of the test potential as a function of excitation energy.

5. Conclusion

Super-heavy elements formed by fusion reaction of heavy ions mainly decay by fission. Measuring the tiny residue cross sections can give a precise information on the fission width and the fission barrier provided we know the fusion probability. Unfortunately it is not the case and one of the main challenges is to find ways to assess the fusion models.[1,4]

An alternative way is to use fission time measurements to locate the super-heavy island of stability.[5] The very long fission times measured by crystal blocking techniques for the $Z = 120$ and 124 nuclei remain unexplained.[7,10]

References

1. C.W. Shen et al, *Sci China Ser G* **10**, 52 (2009)
2. B. Bouriquet et al, *Comp. Phys. Comm.* **159**, 1 (2004)
3. A. Marchix, PhD thesis, Univ. Caen (2007), http://tel.archives-ouvertes.fr/tel-00197012/fr/
4. D. Boilley et al, *CERN-Proceedings* **2010-001**, 479 (2010)
5. M. Morjean et al, *Phys. Rev. Lett.* **101**, 072701 (2008)
6. Y. Abe et al, *Phys. Rep* **275**, 49 (1996)
7. D. Boilley et al, *Int. J. Mod. Phys.* **E17**, 1681 (2008)
8. J.C. Pei et al, *Phys. Rev. Lett.* **102**, 192501 (2009)
9. D. Boilley et al, *Phys. Rev.* **E70**, 056129 (2004)
10. B. Yılmaz et al, in preparation

Conference photo

LIST OF PARTICIPANTS

ADEMARD Guilain
GANIL
Bd. Henri Becquerel
B.P. 55027
F-14076 Caen Cedex 05
France
ademard@ganil.fr

AIT ABDERRAHIM Hamid
SCK•CEN
Boeretang 200
B-2400 Mol
Belgium
haitabde@sckcen.be

AL-ADILI Ali
EC-JRC-IRMM
Retieseweg 111
B-2440 Geel
Belgium
ali.al-adili@ec.europa.eu

ANDREYEV Andrei
University of the West of Scotland
Paisley Campus
Paisley, Renfrewshire PA1 1AG
United Kingdom
Andrei.Andreyev@uws.ac.uk

AYYAD Yassid
Universidad de Santiago de Compostela
Departamento de Fisica de Particulas
Spain
francescyassid.ayyad@usc.es

BAIL Adeline
CEA DAM
DIF
F-91297 Arpajon
France
adeline.bail@cea.fr

BELIER Gilbert
CEA DAM
DIF
F-91297 Arpajon
France
gilbert.belier@cea.fr

BOILLEY David
GANIL
B.P. 55027
F-14076 Caen Cedex 05;
Université de Caen
B.P. 5186
F-14032 Caen Cedex 05
France
Boilley@ganil.fr

CARJAN Nicolae
Université de Bordeaux I
CENBG
F-33175 Gradignan Cedex
France;
NIPNE
Bucharest
Romania
carjan@in2p3.fr

CHATILLON Audrey
CEA DAM
DIF
F-91297 Arpajon
France
audrey.chatillon@cea.fr

CUGNON Joseph
Université de Liège
Dept. AGO
Allée du 6 Août 17, bât. B5
B-4000 Liège 1
Belgium
cugnon@plasma.theo.phys.ulg.ac.be

D'HONDT Pierre
SCK•CEN
Boeretang 200
B-2400 Mol
Belgium
pierre.dhondt@sckcen.be

DORE Diane
CEA Saclay
IRFU/Service de Physique Nucléaire
F-91191 Gif-sur-Yvette
France
diane.dore@cea.fr

FAUST Herbert
Institut Laue-Langevin
6, rue Jules Horowitz
F-38042 Grenoble
France
faust@ill.fr

FURMAN Walter
Joint Institute for Nuclear Research
Joliot Curie 6
R-141980 Dubna
Moscow region
Russia
furman@nf.jinr.ru furman@dubna.ru

GELTENBORT Peter
Institut Laue-Langevin
6, rue Jules Horowitz
F-38042 Grenoble
France
geltenbort@ill.fr

GMUCA Stefan
Institute of Physics
Dubravska cesta 9
SK-84511 Bratislava
Slovakia
gmuca@savba.sk

GOENNENWEIN Friedrich
Universität Tübingen
Auf der Morgenstelle 14
D-72076 Tübingen
Germany
friedrich.goennenwein@uni-tuebingen.de

GOOK Alf
TU Darmstadt
Institut für Kernphysik
Schlossgarten strasse 9
D-64289 Darmstadt
Germany
agook@ikp.tu-darmstadt.de

GOUTTE Héloise
GANIL
B.P. 55027 Caen Cedex 5
France
Heloise.goutte@cea.fr

HAMBSCH Franz-Josef
EC-JRC-IRMM
Retieseweg 111
B-2440 Geel
Belgium
franz-josef.hambsch@ec.europa.eu

HANAPPE Francis
Université Libre de Bruxelles
CP229, av. F.D. Roosevelt 50
B-1050 Brussel
Belgium
fhanappe@ulb.ac.be

HEYSE Jan
SCK•CEN
Boeretang 200
B-2400 Mol
Belgium
jheyse@sckcen.be

HUYSE Mark
Katholieke Universiteit Leuven
Instituut voor Kern- en Stralingsfysica
Celestijnenlaan 200D
B-3001 Leuven
Belgium
Marc.Huyse@fys.kuleuven.ac.be

ITKIS Mikhail
Joint Institute for Nuclear Research
Joliot Curie 6
R-141980 Dubna
Moscow region
Russia
Itkis@flnr.jinr.ru

JURADO Beatriz
CEN Bordeaux-Gradignan
Chemin du Solarium
B.P. 120
F-33175 Gradignan
France
jurado@cenbg.in2p3.fr

LITAIZE Olivier
CEA Cadarache
DEN/DER/SPRC/LEPh, Bat. 230
F-13108 Saint Paul lez Durance
France
Olivier.litaize@cea.fr

MANCUSI Davide
Université de Liège
Fundamental Interactions in Physics and Astrophysics
Allée du 6 Août, 17
B-4000 Liège 1
Belgium
d.mancusi@ulg.ac.be

MUTTERER Manfred
Technische Universität Darmstadt
Institut für Kernphysik
Schlossgartenstrasse 9
D-64289 Darmstadt
Germany
mutterer@email.de

OBERSTEDT Stephan
EC-JRC-IRMM
Retieseweg 111
B-2440 Geel
Belgium
stephan.oberstedt@ec.europa.eu

POLLITT Andrew
The University of Manchester
Room 4.18, Schuster Laboratory
Manchester M13 9PL
United Kingdom
andrew.pollitt@postgrad.manchester.ac.uk

POPESCU Lucia-Ana
SCK•CEN
Boeretang 200
B-2400 Mol
Belgium
lpopescu@sckcen.be

SEROT Olivier
CEA Cadarache
DEN/DER/SPRC/LEPh, Bat. 230
F-13108 Saint Paul lez Durance
France
olivier.serot@cea.fr

OBERSTEDT Andreas
Örebro University
School of Science and Technology
S-70182 Örebro
Sweden
andreas.oberstedt@oru.se

PANOV Igor
Institute for Theoretical and Experimental Physics
B. Cheremuskinskaya st. 25
R-117218 Moscow
Russia
Igor.Panov@itep.ru

POMP Stephan
Uppsala University, Box 516
Department of Physics and Astronomy
Division of Applied Nuclear Physics
S-75120 Uppsala
Sweden
Stephan.Pomp@physics.uu.se

SCHMIDT Karl-Heinz
CENBG/IN2P3
Université Bordeaux I
Chemin du Solarium
B.P. 120
F-33175 Gradignan Cedex
France
schmidt-erzhausen@t-online.de

SIMUTKIN Vasily
Uppsala University, Box 516
Department of Physics and Astronomy
Division of Applied Nuclear Physics
S-75120 Uppsala
Sweden
vasily.simutkin@fysast.uu.se

SMITH Gavin
The University of Manchester
Schuster Laboratory
School of Physics and Astronomy
Manchester M13 9PL
United Kingdom
gavin.smith@manchester.ac.uk

TARRIO Diego
Universidad de Santiago de Compostela
Departamento de Fisica de Particulas
Facultad de Fisica
E-15782 Santiago de Compostela
Spain
diego.tarrio@usc.es

VENHART Martin
Katholieke Universiteit Leuven
Instituut voor Kern- en Stralingsfysica
Celestijnenlaan 200D bus 2418
B-3001 Leuven
Belgium
Martin.Venhart@fys.kuleuven.be

WAGEMANS Cyrillus
Universiteit Gent
Vakgroep Fysica en Sterrenkunde
Proeftuinstraat 86
B-9000 Gent
Belgium
cyrillus.wagemans@ugent.be

ZAGREBAEV Valery
Flerov Laboratory of Nuclear Reactions
Joint Institute for Nuclear Research
Dubna, Moscow Region
Russia
zagrebaev@jinr.ru

STUTTGE Louise
IPHC-DRS
23, rue du Loess
B.P. 28
F-67037 Strasbourg Cedex 2
France
stuttge@in2p3.fr

TSEKHANOVICH Igor
The University of Manchester
Department of Physics and Astronomy
M13 9PL Manchester
United Kingdom
igor.tsekhanovich@manchester.ac.uk

VERMOTE Sofie
Universiteit Gent
Vakgroep Fysica en Sterrenkunde
Proeftuinstraat 86
B-9000 Gent
Belgium
sofie.vermote@ugent.be

WAGEMANS Jan
SCK•CEN
Boeretang 200
B-2400 Mol
Belgium
jan.wagemans@sckcen.be

AUTHOR INDEX

Abe, Y. 297
Ademard, G. 131
Aït Abderrahim, H. 191
Al-Adili, A. 99
Almahamid, I. 145
Álvarez, H. 267
Andersson, P. 107
Audouin, L. 199, 263
Aupiais, J. 231
Ayyad, Y. 267

Bacquias, A. 267
Bail, A. 199
Barday, R. 123
Belgya, T. 207, 223
Belier, G. 173, 199, 231
Benlliure, J. 199, 267
Bernard, R. 45
Bevilacqua, R. 107
Billnert, R. 223
Blomgren, J. 107
Bogachev, A. A. 273
Boilley, D. 297
Bonnet, E. 131
Borcea, R. 207
Boudard, A. 267

Cano-ott, D. 223
Cârjan, N. 31
Casarejos, E. 267
Charity, R. 135

Chatillon, A. 173, 199
Chbihi, A. 131
Chernykh, M. 123
Chernysheva, E. 181
Cugnon, J. 135

Dare, J. A. 115, 215
Devlin, M. 173
Dore, D. 199
Dorvaux, O. 181
Durán, I. 263

Eckardt, C. 123
Enders, J. 123
Enqvist, T. 267

Faust, H. 89
Föhr, V. 267
Frankland, J. D. 131
Furman, W. 53

Geltenbort, P. 145
Gönnenwein, F. 3, 181
Göök, A. 123, 207, 223
Goutte, H. 45
Granier, T. 173
Greiner, W. 289

Haight, R. C. 173
Hambsch, F.-J. 99, 123, 165, 181, 207, 223

Hanappe, F. 181
Heyse, J. 145, 191, 239

Itkis, I. M. 273
Itkis, M. G. 273

Jurado, B. 73, 81, 199

Karlsson, J. 223
Karpov, A. V. 289
Kelic, A. 267
Kezzar, K. 267
Kis, Z. 207, 223
Kknyazheva, G. N. 273
Kopatch, Yu. 181
Korneev, I. Yu. 255
Kozulin, E. M. 273

Lallouet, Y. 297
Laurent, B. 173, 199
Ledoux, X. 223
Litaize, O. 65

Mancusi, D. 135
Marmouget, J.-G. 223
Martinez-Perez, T. 207, 223
Meulders, J. 107
Mutterer, M. 155, 181

Nelson, R .O. 173
Noda, S. 173

O'Donnell, J. M. 173
Oberstedt, A. 123, 207, 223
Oberstedt, S. 99, 123, 165, 207, 223
Onegin, M. S. 107
Österlund, M. 107

Panebianco, S. 199
Panov, I. V. 255
Paradela, C. 263, 267
Pérez-Loureiro, D. 267
Pleskac, R. 267
Pollitt, A. J. 115, 215
Poltoratska, Y. 123
Pomp, S. 99, 107
Popescu, L. 239
Prieels, P. R. 107

Rao, L. 145
Rauscher, T. 255
Reymund, F. 199
Richter, A. 123
Ryzhov, I. V. 107

S. Leray, 267
Schmidt, K.-H. 73, 81, 199
Serot, O. 65, 145, 247
Simutkin, V. D. 107
Smith, A. G. 115, 215
Soldner, T. 145
Stuttgé, L. 181
Szentmiklosi, L. 207, 223

Taieb, J. 173, 199
Takács, K. 207, 223
Tarrio, D. 263, 267
Tassan-Got, L. 199, 263
Thielemann, F.-K. 255
Tian, G. 145
Tsekhanovich, I. 115, 215
Tutin, G. A. 107

Vaishnene, L. A. 107
Van Gils, J. 247
Vermote, S. 247, 145
Von Neumann-Cosel, P. 123

Wagemans, C.   145, 239, 247
Wagemans, J.   191, 239
Wagner, M.   123
Wieleczko, J. P.   131

Yilmaz, B.   297

Zagrebaev, V. I.   289
Zeynalov, Sh.   165, 207